Privacy in Vehicular Networks

In an era where vehicular networks and Location-Based Services (LBS) are rapidly expanding, safeguarding location privacy has become a critical challenge. *Privacy in Vehicular Networks* delves into the complexities of protecting sensitive location data within the dynamic and decentralized environment of vehicular networks. This book stands out by addressing both the theoretical and practical aspects of location privacy, offering a thorough analysis of existing vulnerabilities and innovative solutions.

This book meticulously examines the interplay between location privacy and the operational necessities of road networks. It introduces a differential privacy framework tailored specifically for vehicular environments, ensuring robust protection against various types of privacy breaches. By integrating advanced detection algorithms and personalized obfuscation schemes, the book provides a multi-faceted approach to enhancing location privacy without compromising data utility.

The key features of this book can be summarized as follows:

- **Comprehensive Analysis:** Detailed examination of location privacy requirements and existing preservation mechanisms.
- **Innovative Solutions:** Introduction of a Personalized Location Privacy-Preserving (PLPP) mechanism based on Road Network-Indistinguishability (RN-I).
- **Advanced Detection:** Utilization of Convolutional Neural Networks (CNN) for detecting illegal trajectories and enhancing data integrity.
- **Collective Security:** Implementation of the Cloaking Region Obfuscation (CRO) mechanism to secure multiple vehicles in high-density road networks.
- **Holistic Approach:** Joint Trajectory Obfuscation and Pseudonym Swapping (JTOPS) mechanism to seamlessly integrate privacy preservation with traffic management.
- **Future-Ready:** Exploration of upcoming challenges and recommendations for future research in vehicular network privacy.

This book is essential for researchers, practitioners, and policymakers in the fields of vehicular networks, data privacy, and cybersecurity. It provides valuable insights for anyone involved in the development and implementation of LBS, ensuring they are equipped with the knowledge to protect user privacy effectively.

Baihe Ma received his B.E. and M.E. degrees from Xidian University, China, in 2016 and 2019, respectively, and he received his Ph.D. degree from the University of Technology Sydney, Australia, in 2024. His primary research interests encompass cybersecurity, data privacy, location privacy, and trust management. He has published 26 papers and 5 patents.

Dr. Xu Wang received his B.E. from Beijing Information Science and Technology University, China, in 2010, and dual Ph.D. degrees from Beijing University of Posts and Telecommunications, China, in 2019 and the University of Technology Sydney, Australia, in 2020. He is currently a Senior Lecturer in the Faculty of Engineering and IT (FEIT) at UTS and teaches Cybersecurity subjects. His primary research interests encompass several domains, including blockchain, cybersecurity, vehicle privacy and security, IoT, and AI. He has published 42 research papers and 3 patents.

Dr. Wei Ni received the B.E. and Ph.D. degrees in Communication Science and Engineering from Fudan University, Shanghai, China, in 2000 and 2005, respectively. He is a Principal Research Scientist at CSIRO, Sydney, Australia, a Fellow of IEEE, a Conjoint Professor at the University of New South Wales, an Adjunct Professor at the University of Technology Sydney, and an Honorary Professor at Macquarie University. He was a Postdoctoral Fellow at Shanghai Jiaotong University from 2005 to 2008; Deputy Project Manager at Bell Labs, Alcatel/Alcatel-Lucent from 2005 to 2008; and Senior Researcher at Nokia from 2008 to 2009. He has authored eight book chapters, more than 300 journal papers, more than 100 conference papers, 26 patents, and ten standard proposals accepted by IEEE. His research interests include machine learning, online learning, and stochastic optimization and their applications to the security, integrity, and efficiency of network systems. Dr. Ni has served as an Editor for IEEE Transactions on Vehicular Technology since 2022, IEEE Transactions on Wireless Communications since 2018, and Cambridge Press New Research Directions: Cyber-Physical Systems since 2022. He served as the Chair of the IEEE Vehicular Technology Society New South Wales Chapter from 2020 to 2022, the Secretary and then the Vice-Chair of the Chapter from 2015 to 2019, Track Chair for VTC-Spring 2017, Track Co-chair for IEEE VTC-Spring 2016, Publication Chair for BodyNet 2015, and Student Travel Grant Chair for WPMC 2014.

Ren Ping Liu received his B.E. degree from Beijing University of Posts and Telecommunications, China, and the Ph.D. degree from the University of Newcastle, Australia, in 1985 and 1996, respectively. Ren Ping Liu is a Professor and Head of the Discipline of Network & Cybersecurity at the University of Technology Sydney (UTS). As a research leader, a certified network professional, and a full-stack web developer, he has delivered networking and cybersecurity solutions to government agencies and industry customers. His research interests include wireless networking, 5G, IoT, Vehicular Networks, 6G, Cybersecurity, and Blockchain. He has supervised over 30 Ph.D. students and has over 200 research publications. Professor Liu was the winner of NSW iAwards 2020 for leading the BeFAQT (Blockchain-enabled Fish provenance And Quality Tracking) project. He was awarded the Australian Engineering Innovation Award 2012 and the CSIRO Chairman's Medal for his contribution to the Wireless Backhaul project. Professor Liu was the founding chair of the IEEE NSW VTS Chapter and a Senior Member of IEEE.

1. Baihe Ma: Google Scholar
   https://scholar.google.com/citations?user=AfvXcVoAAAAJ&hl=en&oi=ao
2. Xu Wang Profile
   https://profiles.uts.edu.au/Xu.Wang
3. Wei Ni Profile
   https://people.csiro.au/n/w/wei-ni
4. Ren Ping Liu Profile
   https://profiles.uts.edu.au/renping.liu

Privacy in Vehicular Networks

Challenges and Solutions

Baihe Ma, Xu Wang, Wei Ni, and Ren Ping Liu

CRC Press
Taylor & Francis Group
Boca Raton London New York

CRC Press is an imprint of the
Taylor & Francis Group, an informa business

Designed cover image: © Shutterstock

First edition published 2025
by CRC Press
2385 NW Executive Center Drive, Suite 320, Boca Raton FL 33431

and by CRC Press
4 Park Square, Milton Park, Abingdon, Oxon, OX14 4RN

CRC Press is an imprint of Taylor & Francis Group, LLC

Publisher's Note: This book has been prepared from camera-ready copy provided by authors.

ISBN: 978-1-032-82352-2 (hbk)
ISBN: 978-1-032-85417-5 (pbk)
ISBN: 978-1-003-51804-4 (ebk)

DOI: 10.1201/9781003518044

Typeset in CMR10
by KnowledgeWorks Global Ltd.

Contents

Abstract

The rapid expansion of Location-Based Services (LBS) within the Internet of Vehicles (IoV) poses significant privacy risks to vehicular networks, where sensitive location data and driver information are susceptible to exposure. This book, documented across several chapters, introduces a differential privacy framework tailored for vehicular environments, addressing the complex interplay between location privacy and the operational necessities of road networks.

In Chapters 1 and 2, this book presents a detailed illustration of the location privacy requirements in the era and the existing location privacy-preserving mechanisms.

Chapter 3 introduces a Personalized Location Privacy-Preserving (PLPP) mechanism, grounded in the concept of Road Network-Indistinguishability (RN-I), which dynamically and adaptively protects individual driver privacy without compromising data utility.

In Chapter 4, we further present an approach to trajectory data, identifying and mitigating illegal trajectories using advanced detection algorithms, ensuring the integrity and utility of shared data in the IoV. This includes the Trajectory-Indistinguishability (T-I) metric and real-time location privacy protection.

Chapter 5 discusses the Cloaking Region Obfuscation (CRO) mechanism, leveraging differential privacy to secure multiple vehicles collectively, a critical innovation given the high-density nature of road networks.

A Joint Trajectory Obfuscation and Pseudonym Swapping (JTOPS) mechanism is detailed in Chapter 6, which seamlessly integrates privacy preservation with essential traffic management functionalities.

Concluding in Chapter 7, this book not only enhances current understanding and capabilities in vehicular network privacy but also sets the stage for future explorations into more resilient, adaptive privacy mechanisms suitable for the evolving landscape of vehicular communications and its privacy challenges.

List of Figures

List of Tables

1

Introduction

The rapid development of vehicular networks has propelled the widespread use of Location-Based Services (LBSs), which rely heavily on precise location data. While these services offer significant benefits, such as enhanced navigation and real-time traffic updates, they also pose serious threats to location privacy. This thesis addresses the vulnerabilities inherent in current vehicular networks where location data can be exploited by adversaries to infringe on privacy. It investigates the challenges posed by the dynamic, highly mobile, and decentralized nature of vehicular networks that complicate the application of traditional Location Privacy-Preserving Mechanisms (LPPMs).

This chapter first highlights the importance of location data and location privacy concerns. Then, the characteristics of vehicular networks and location privacy requirements are provided. Finally, this chapter summarizes the research gap in existing works and the structure of this book.

1.1 Location Data and Location Privacy Concerns

Being an integral part of mobile systems, e.g., the fifth- and sixth-generation (5/6G) systems, vehicular networks are required to provide credible information for driving assistance [155]. Over the past decades, multifarious applications have been developed for vehicular networks to offer various Location-Based Services LBSs. LBSs can be classified by functions into navigation, weather, venue finders, social media, and crowd-sensing [312]. By reporting location data to the LBSs, the drivers can search their destinations, check traffic conditions, and view the weather [116].

The development of LBS leads to rapidly increasing vehicular applications related to location data [391]. This poses threats to the location privacy of drivers, as LBS servers can learn the geographical locations of drivers [489]. Adversaries can collude with untrusted LBS servers to infer drivers' personal information from the location data [211]. The adversaries may also compromise trusted LBS servers or eavesdrop on the communications between the drivers and LBS servers for the location data [250, 485]. Compared with general mobile LBS users, vehicles can be relatively easily tracked because they only travel on roads and follow traffic rules [28, 346].

With increasingly popular communication sensing techniques, including cameras and vehicle-to-vehicle communications, a lot of location data will be generated in vehicular networks, in real-time and on a large scale. Location data shared in vehicular networks have characteristics such as massive, highly correlated, dynamic, and unequal importance [234]. Location data is important to help improve the safety and efficiency of vehicular networks, including vehicle infrastructure and pedestrians.

DOI: 10.1201/9781003518044-1

• *Massive Data:* There are enormous amounts of location information when the drivers apply LBSs [421].

• *High Correlation:* Location data is correlated, which can disclose other information when exposed.

• *Dynamic Topology:* Location data frequently changes over time.

• *Uneven Significance:* Different location information has various significance to various drivers.

Driver's home and workplace addresses tend to be more important than his shopping locations.

• *Driver Safety:* One of the most important purposes of location data transmitted in vehicular networks is safe driving [282]. The data transmission in vehicular networks should primarily satisfy the driving requirements [262]. For example, the transmission should ensure a high data utility of location data after being protected for location-sensitive LBSs, e.g., navigation.

Sharing the location data in LBSs of vehicular networks raises privacy issues [255]. LBSs in vehicular networks ask drivers to expose their locations for work as planned, but untrusted LBSs and potential adversaries can lead to the leakage of the location data [397]. Adversaries can collude with untrusted LBS servers, attack trusted LBS servers, and eavesdrop on communications channels to gain the shared location data [255]. This poses the location data at risk of being obtained by adversaries. Adversaries can threaten the drivers' privacy by analyzing the shared location data.

• *Exposure of Driver's Private Information:* Adversaries can infer private information by analyzing the location data transmitted in vehicular networks, such as home addresses, religions, political parties, jobs, and work addresses [251, 46]. For example, a driver's religion can be estimated if the driver drives to religious buildings periodically. The inferred private information can be utilized to build models of targets for realizing attacks [493]. Adversaries can attack a specific driver, threaten a group of targets, or even sabotage a whole system [385].

• *Physical Attack:* The disclosure of the location privacy can lead to dangerous physical attacks, such as stalking, mugging, or burglary [266]. For example, attackers can jam the traffic and induce the driver's trajectory[1].

• *Exposure of Other Vehicles' Private Information:* The drivers' trajectories can also be utilized to disclose the privacy of other drivers and predict other drivers' mobility [469]. For example, adversaries can analyze the driver's encounter history to obtain the trajectories of other drivers.

The advanced techniques in vehicular networks introduce new threats to drivers' location privacy. With the development of vehicular networks, adversaries have a high probability of launching attacks across multiple vehicular network layers. When facing cross-layer attacks, location privacy protection would be increasingly difficult with the existing LPPMs. For example, the Pegasus (spyware) developed in Israel can be utilized to obtain the driver's historical encounter messages [207]. Adversaries can minimize the estimation error to achieve high-precision localization for the target driver by analyzing the historical encounter messages and the eavesdropped messages. The development of communication capabilities allows the vehicles to share data frequently with other entities in future vehicular networks, leading to the current LPPMs, which focus on a single layer, not being able to protect location privacy when the adversaries launch attacks across multiple layers. The existing LPPMs should be improved to meet the privacy requirements in complex composite scenarios.

[1] `https://simonweckert.com/googlemapshacks.html`

1.2 Location Privacy Requirement

Researchers have different definitions of location privacy [66, 142, 232]. In this thesis, location privacy can be defined as a branch of information privacy that deals with the location data to assist the drivers in deciding on when, what, and how to share such data [99]. The data shared in the vehicular networks should not threaten the driver's security [153]. The privacy requirements in vehicular networks can be classified as follows [15, 17].

- *Confidentiality:* The communication data should be well protected when a vehicle applies LBSs. The data cannot be exposed to unauthenticated parties [328].
- *Anonymity:* By analyzing the data from the same identity, the adversary can launch attacks in a long-period window. Hence, the identical information of the vehicles in the released data should be protected to avoid disclosure of the actual identity [328].
- *Unlinkability:* The link between the data and the identity to keep it safe from the adversary and unauthorized third parties [302].
- *Contextual unobservability:* The released data from the same vehicle do not breach the privacy of each other [25].

1.2.1 Characteristic of Vehicular Networks

The characteristics of vehicular networks, such as unlimited transmission power, higher computational capabilities, and predictable mobility, differ from those of other scenarios [461]. The characteristics of current vehicular networks are classified into topology features, node features, and transmission features. As shown in Figure 1.1, the following are the unique characteristics of vehicular networks, which are important to the location privacy in the study.

Topology Feature: The topology features of vehicular networks related to location privacy include mobility, dynamic network topology, real-time constraints, frequent network disconnection, and volatility.

- *Mobility:* A slight delay in V2X communication can lead to severe problems since vehicles in vehicular networks move at high speeds [468]. It is impossible to employ traditional LPPMs, e.g., handshake-based authentication technologies, since most of the encountered vehicles only communicate once and do not have enough time for handshake message authentication [387]. In vehicular networks, vehicles frequently change their points of network attachment frequently [61]. Thus, the LPPMs in vehicular networks should have mobility management methods to meet the requirements, such as seamless mobility and scalability [86].
- *Dynamic Network Topology:* The topology of vehicular networks changes quickly so that the communication duration of LPPMs is limited [96]. The density of entities on the road networks is frequently changed due to the dynamic network topology of vehicular networks, so communication recourse could be unevenly distributed [26]. For example, cooperation-based LPPMs (e.g., pseudonym swap) could perform well in the city scenario rather than in the suburb scenario, as the number of nodes on the road networks in a city is much more than that in the suburbs.
- *Real-Time Constraint:* The data transmission in vehicular networks is time-limited [405], and LBSs in vehicular networks (e.g., accident warning information) need time-critical messages [73]. Thus, one of the critical requirements in vehicular networks is that LPPMs should allow messages to be transmitted within an acceptable period. Nevertheless, verifying a time-critical message is difficult because the authentication process can increase the time delay [93].

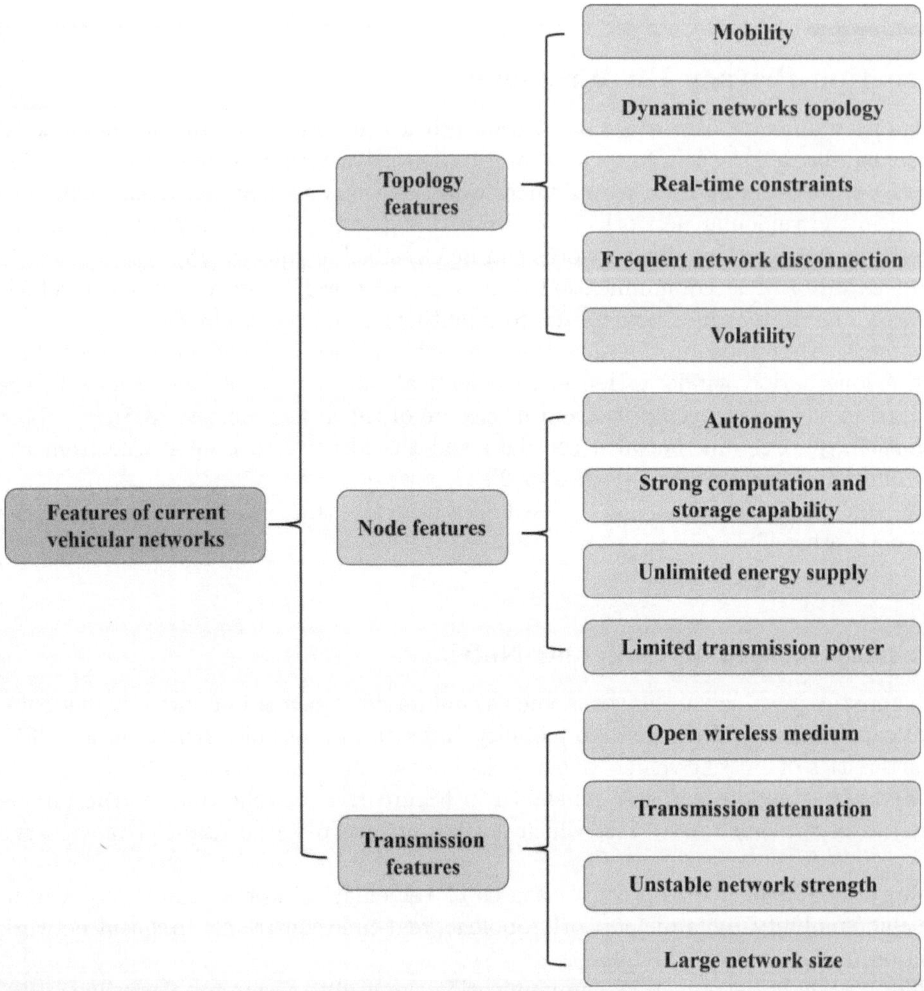

FIGURE 1.1: Features of vehicular networks can be divided into three parts, i.e., topology features, node features, and transmission features.

• *Frequent Network Disconnection:* Vehicles frequently disconnect networks due to the high-speed movement of the vehicles and the influence of the environment [72]. A great number of vehicles in the same region can also lead to server disconnection [279]. Thus, the LPPMs should be robust to provide location privacy protection in such a scenario.

• *Connection Volatility:* The connections among vehicles are random because of the high mobility level [323]. The connectivity between two vehicles could be lost easily, and connections may remain within a specific wireless hop in a short period of time [372]. Vehicular networks lack a long-lived context due to the random and short connection period. Thus, it is almost impossible to utilize long-live password-based LPPMs [113].

Node Feature: The features of nodes in vehicular networks related to location privacy are autonomy, strong computation and storage capability, unlimited energy supply, and limited transmission power.

• *Autonomy:* The vehicles in vehicular networks have the authority to send, route, and receive data with limited control of centralized authority [115]. On-Board Units (OBUs) and RSUs can process the data independently [186]. Hence, decentralized LPPMs are fully considered in vehicular networks.

• *Strong Computation and Storage Capability:* The nodes in vehicular networks are required to process an extensive amount of data among vehicles and infrastructures [309]. The computation and storage capability of nodes in vehicular networks is high, so the computing resource of LPPMs could be unlimited [189].

• *Unlimited Energy Supply:* The energy consumption is unlimited in vehicular networks [177]. OBUs are supplied by vehicle batteries, which satisfy the task operations [126]. Therefore, LPPMs in vehicular networks could ignore the energy limitation.

• *Limited Transmission Power:* The transmission power is limited in vehicular networks due to the characteristics of the communication protocols and the wireless access of the vehicular environment [2]. The covered area of each node in vehicular networks is limited by the transmission power. The communication region of decentralized LPPMs in vehicular networks is limited.

Transmission Features: The transmission features of vehicular networks related to location privacy are open wireless medium, transmission attenuation, unstable network strength, and large network size.

• *Open Wireless Medium:* In vehicular networks, the transmission medium is the air. Security is an important issue in vehicular networks due to the characteristics of the open wireless medium [198]. Attacks targeting open wireless mediums can also be launched to threaten location privacy in vehicular networks.

• *Transmission Attenuation:* The features of digital transmission in vehicular networks include diffraction, reflection, dispersion, refraction, and scattering [284]. These features lead to several limitations for the dedicated short-range communication in vehicular networks [376]. These features could expose the driver's location data even though the location data is well protected by LPPMs [326].

• *Unstable Network Capability:* The communication and computation capabilities of vehicle networks are unstable and affected by real-time traffic conditions [480]. For example, the network strength among vehicles can be very high in a traffic jam, as a large number of vehicles could stay in the same region and form the vehicular network. Thus, the privacy-protection capability of cooperation-based LPPMs (e.g., pseudonym swap) could be unstable.

• *Large Network Size:* Vehicular networks can cover a large area, especially downtown or on highways. However, the communication region of vehicles in vehicular networks is limited. Thus, the LPPMs should consider both the large size of vehicular networks and the small size of the communication regions.

1.2.2 Application Scenario of LPPM

The location data is protected by real-time or batch LPPMs during the collection phase and protected by offline LPPMs at the publication phase, as shown in Figure 1.2. The data is protected by offline LPPMs if it is processed through the server to the publication stage [174], while the data is protected by real-time LPPMs before being sent to time-sensitive LBSs [286]. The batch-wise LPPMs protect location data when multiple drivers aggregate and send location data to time-insensitive LBSs [398]. The offline LPPMs are employed when LBSs publish the trajectory data for analysis to protect the whole dataset rather than the real-time location data [53].

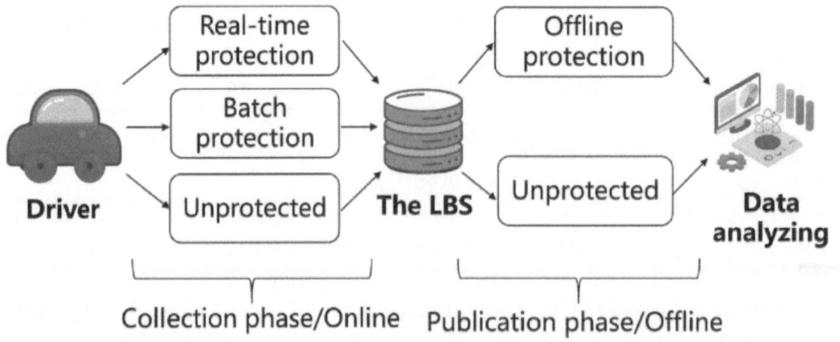

FIGURE 1.2: The cases of employing LPPM to protect location privacy.

1.2.3 Key Performance Index of LPPM

The existing works define three major metrics to evaluate and compare the performances of different LPPMs as follows [312, 196, 343]:

• *Privacy:* The privacy metric measures the location-privacy protection capability of the LPPMs under the adversaries' attacks [91, 266, 255].

• *Data Utility:* The data utility metric is determined by the LBSs' and drivers' requirements. The privacy and data utility metrics are conflicting [196, 168]. For example, the data utility will achieve its peak when the location information is unprotected, and vice versa.

• *Efficiency:* The efficiency metric measures the time and storage cost of the LPPMs [321], which uses computational overhead, storage overhead, scalability, and tolerance of error as metrics [431, 444].

1.3 Gap of Existing Works

Obfuscation-based schemes can protect location privacy against untrusted LBS servers and eavesdroppers [219, 388, 31]. The idea is based on the local differential privacy technology where users locally perturb raw data with differential privacy schemes and employ the obfuscated data rather than the accurate data [402]. In obfuscation-based schemes, drivers generate obfuscated locations by adding controllable noise in their actual locations and report the obfuscated locations in LBS requests [321]. The obfuscated locations are selected according to the distances between the obfuscated and actual locations under the constraint of indistinguishability [31]. Although the obfuscation-based schemes compromise location accuracy, they can still be used in location-insensitive LBS, such as location-based recommendations [271].

Obfuscation-based schemes have not been rigorously developed for vehicles in road networks. In general-purpose obfuscation schemes, e.g., two-dimensional (2D) Laplace location privacy-preserving schemes, the distance between an actual location and the corresponding obfuscated location is measured using the Euclidean distance [31, 166, 388]. This, however, underestimates the distance between two locations on road networks because the route distance is no shorter than the Euclidean distance between any two locations [33]. Another issue of general-purpose obfuscation schemes is that they may generate off-road obfuscated locations, for example, on a river [381]. Additional steps can be introduced to avoid off-road locations, which, however, increase the computational complexity [184].

The privacy-preserving levels of locations are expected to be fine-tuned across road networks to balance data utility and location-privacy protection [471]. Drivers can be sensitive to a small number of locations in road networks. If all locations are obfuscated at the highest level as the sensitive locations, the insensitive locations are over-obfuscated, and the data utility is penalized [414]. If only the sensitive locations are obfuscated, the sensitive locations are statistically different from the insensitive locations. This compromises location privacy [3].

In existing obfuscation-based mechanisms, malicious drivers breach legal drivers' profit. For example, malicious drivers can occupy more benefits than they deserve by deliberately modifying their trajectory data in Taxi service [4, 132]. Malicious drivers also use location privacy-preserving schemes to protect their location data. By analyzing location data, which concludes illegal location data, smart city applications cannot provide an acceptable QoS of LBS. Therefore, LBS should detect illegal location data to ensure high QoS. If the malicious drivers employ location privacy-preserving schemes (e.g., obfuscation schemes) as the legal drivers, detecting illegal data becomes difficult.

The temporal information has not been well considered in road networks by the existing obfuscation mechanisms [106]. If an adversary can track the IDentities (ID) of the vehicles, it can reduce the perturbation noise by linking the communications of a specific vehicle ID over a long period, such as a long-observation attack [289]. As a result, the existing obfuscation-based mechanisms can hardly protect location privacy on a long-term basis [50].

1.4 Structure of This Book

This book aims to bridge the gaps between the existing LPPMs by focusing on real-world road networks.

Chapter 2 provides a comprehensive review of existing LPPMs relevant to this thesis. We first illustrate the existing localization techniques which introduce challenges to the existing LPPMs. We then review the existing LPPMs by comparing their strengths and limitations. Finally, we discuss how LPPMs balance location privacy and data utility in vehicular networks.

In Chapter 3, we define a new measure of location indistinguishability for road networks by applying the concept of differential privacy. The indistinguishability is measured only on road networks (as opposed to continuous 2D space). Route distances (i.e., distances along roads) are used to derive the differential privacy upper bound. Off-road locations are precluded from obfuscation operations. We introduce a new dual-obfuscation algorithm that first probabilistically obfuscates an actual location to a connection and then obfuscates the location into a road interval between the actual location and the obfuscated connection. By carefully designing the probabilities, we prove that the dual-obfuscation design satisfies the new differential privacy-based definition of road network-indistinguishability. Considering the non-uniform location privacy requirements of a driver at different locations, we apply the nearest neighbor interpolation to specify the privacy budgets for all locations based on the sensitive locations of the driver. The vehicle locations can be obfuscated consistently without exposing the sensitive locations while improving the utility of the location data of less sensitive locations. We conduct comprehensive experiments on two real-world trajectory datasets and compare the presented PLPP scheme with the existing 2D Laplace location privacy-preserving schemes [388, 166, 12]. Experimental results show that the PLPP scheme outperforms the 2D Laplace schemes in terms of data utility, privacy-preserving level, and efficiency.

In Chapter 4, we research the mechanisms that improve data utility and privacy protection in road networks. We employ differential privacy in the personalized obfuscation scheme to adaptively protect drivers' location privacy and provide high Quality of Services (QoS) of LBS in road networks. We present a convolutional neural network (CNN)-based detection mechanism to detect illegal trajectories without requiring the drivers' actual location. The scheme has high accuracy in detecting illegal trajectories even if the drivers protect actual location data with various noise sizes. We further protect location privacy by cooperating with multiple vehicles in road networks. Firstly, we develop differential privacy in road networks and extend the introduced RN-Indistinguishability that quantifies road network location privacy with unique road network features. Using the extended RN-indistinguishability, we design the CRO mechanism that employs the route distances to quantify the indistinguishability of road locations. We prove that the CRO mechanism satisfies the RN-indistinguishability. The CRO mechanism can be extended with general road network features without breaching differential privacy. Then, we use pseudonym swapping to incorporate multiple vehicles in the time and spatial domains. We demonstrate that pseudonym swaps can be considered a differential privacy process for any two identities sharing the same pseudonym-swapping candidate set. We analytically prove that jointly using pseudonym swapping and obfuscation in vehicular networks can achieve higher privacy protection than separately using them.

Chapter 5 introduces a new unified privacy-preserving measure, i.e., T-I, treating trajectory obfuscation and pseudonym-based mechanisms as a holistic process. T-I extends the applicability of ε-differential privacy (DP) by quantifying the distinguishability between vehicles based on their historical information in the time domain and current information in the spatial domain. Building upon T-I, we present a novel Joint Trajectory Obfuscation and Pseudonym Swapping (JTOPS) mechanism joining pseudonym swapping and trajectory obfuscation with a crafted criteria. The mechanism is proven to combine two differential privacy processes, without introducing the additivity composition theorem of ε-DP. The pseudonym swapping does not require vehicles to disclose their private data, as the required parameters are computed locally and then transmitted to the coordinator. Thus, the knowledge of the coordinator, which assists in pseudonym swapping, is limited, ensuring the introduced pseudonym-swapping process can effectively resist collusion attacks.

Chapter 6 concludes the thesis and outlines our future works, while Chapter 7 lists the publications of the author.

2

Earlier Development of LPPM

In this chapter, the existing tracking techniques in the different scenarios are first summarized from the aspect of location privacy, as shown in Table 2.1. Then, we illustrate the localization requirements, list the general adversary models, and provide possible privacy threats in current vehicular networks. The adversaries that have been well-considered are presented by highlighting the threats introduced by localization techniques. Then, we assess the existing LPPMs in three categories and discuss the limitations of each category, as shown in Table 2.2.

The existing LPPMs are divided into user-side, server-side, and user-server-interface LPPMs according to the phase that the LPPMs are allocated, by considering the threats of localization techniques [183]. The user-side LPPMs allow the drivers to process data before sending it to the LBSs, while the server-side LPPMs process the aggregated location data in the dataset. The user-server-interface LPPMs use trusted third parties and secure communication to ensure that the location data is secure in transmission. We summarize the limitations of LPPMs from the aspect of localization and communication requirements, and we review the method to balance location privacy and data utility from the aspects of theory and practice. The theoretical methods are classified into blockchain, adaptive parameters, hybrid, encryption, element simplification, and virtual nodes, where we provide existing works to illustrate the method in detail. The real-world in-vehicle tracking techniques, i.e., COVID-19 tracking applications, are used as examples to discuss the balance between location privacy and quality of service in practice.

DOI: 10.1201/9781003518044-2

TABLE 2.1: Existing localization techniques and their challenges.

Category	Technique	Demand	Description	Challenge
Sensing infrastruct-ure-based tracking	Magnetic sensors [121, 122, 378, 379, 478]	Obtain magnetic induction signal generated by the vehicles	Calculate three-dimensional location and two-dimensional orientation information of the vehicles	Malicious sensors are difficult to be detected; The adversaries can collude with sensors; The adversaries can hijack legal sensors; No existing work can defend the adversaries with sensors
	Inductive loop detectors [136, 335]	The inductive loop generated by the vehicles	Can be deployed at intersections on road networks; Can monitor information like speed, volume, and size	
	Beacon message-based sensors [188, 283, 374]	Beacon messages like AVI tags, RFID tags, GPS, and MAC	Extract private information from beacon messages	
		Use ML and AI to analyze the geometrical	Observe the physical features of	It is impossible to hide the physical features of the vehicles; Image iden-tification has a high accuracy;

continued on next page

continued from previous page

Optical vision-based tracking	ML and AI [22, 119, 447]	information of the captured camera frames	the vehicles; Image identification background	The adversaries can directly observe the vehicles in the real world; No existing LPPM can defend the adversaries in such a scenario.
Vehicle driving log-based tracking	Reverse engineering [308, 210, 108]	GPS and CAN data	Read CAN data from OBD and use reverse engineering technique to calculate location data	Multiple OBD readers can be combined for localization
	In-vehicle communication [71, 275, 337, 425, 298, 151, 227, 140, 141]	In-vehicle communication system; Drivers' mobiles	Hijack in-vehicle communication system to obtain location data; Hijack mobile phones to read location data	Few existing LPPMs focus on hardware layer location privacy; An in-vehicle communication system is difficult to be considered in LPPMs for LBSs; The mobile devices are almost impossible to be

continued on next page

continued from previous page

				isolated in the vehicles
Cellular radio-based tracking	Antenna localization [314, 65, 112, 159]	Base stations Antenna signal	Use the direction of the antenna signal for localization	Antennas are equipped in the vehicles for communication; Signals can be detected in the physical world; Based station is necessary for communication; Localization with multiple based stations is very easy; No existing works can defend cellular radio-based localization
Upper-layer message--based tracking vehicular networks	Estimation actual information [40, 340, 169, 59, 56, 417, 57]	Messages in vehicular networks	The adversaries can launch multiple attacks to obtain messages that are transmitted in vehicular networks, with which	Various attacks can be utilized. The adversaries can launch attacks that cross multiple

continued on next page

continued from previous page				
			it can infer the drivers' actual locations	layers

2.1 Localization and Tracking Techniques in Vehicular Networks

This section provides a comprehensive illustration of the existing localization and tracking techniques and the existing location privacy threat in vehicular networks.

2.1.1 Localization and Tracking Techniques

Over the past decades, tracking techniques have been presented for high-precision LBSs and anti-theft systems [89]. However, the adversaries can use tracking techniques that branch the drivers' location privacy. Multiple LPPMs are presented to prevent the adversaries from obtaining the drivers' privacy.

The existing tracking techniques can be classified as sensing infrastructure-based, optical vision-based, vehicle driving log-based, cellular radio-based, and upper-layer message-based. The adversaries with upper-layer message-based tracking techniques are well considered in the existing LPPMs, while the adversaries with vehicle-driving log-based tracking techniques are overlooked by the existing studies. Location privacy under sensing infrastructure-based, optical vision-based, and cellular radio-based tracking is difficult to protect.

2.1.1.1 Sensing Infrastructure-Based Tracking

Sensing infrastructure-based tracking techniques detect vehicle trajectories by using equipment such as inductive loop, infrared, ultrasonic, microwave, magnetic and piezoelectric sensors [470]. The driving features captured by the sensors can be utilized for tracking.

• *Magnetic Sensor:* Vehicles disrupt the Earth's magnetic field that generates magnetic induction signals [192]. The magnetic sensors can expose the 3D location and 2D orientation information of the vehicles [379]. The magnetic sensors can capture the generated signals to analyze the travel time and vehicles' identifications [122]. The existing works [121, 122, 478, 378] modeled the magnetic field perturbations caused by the vehicles and extracted the magnetic waveforms of the vehicles. With the models and the waveforms, the techniques can track the vehicles and infer their driving status.

• *Inductive Loop Detector:* The inductive loop detectors are widely employed to obtain traffic data, which can be deployed at intersections on road networks to monitor traffic [195]. The inductive loop detection can provide vehicle information, such as speed, volume, and size [192]. The existing works [335, 136] used the data obtained from inductive loop detectors to increase the tracking accuracy.

• *Beacon-Message-Based Sensor:* The Beacon-message-based tracking techniques utilize messages like Automatic Vehicle Identification (AVI) tags and Radio Frequency IDentification (RFID) tags, Global Positioning Systems (GPS), and Medium Access Control (MAC) [324, 171, 278, 477]. The existing works [374, 283, 188] tracked vehicles with multiple beacon-message-based sensors because a single sensor is susceptible to the vehicles' driving status and environment.

No existing LPPMs can protect drivers' location privacy under the sensing of adversaries who utilize multiple sensors. The existing studies focused on protecting location data captured by legal sensors through techniques such as blockchain, obfuscation, and anonymity.

2.1.1.2 Optical Vision-Based Tracking

The optical vision-based tracking estimates vehicles' locations by analyzing the geometrical information of the captured camera frames (i.e., image pixels) [22]. The optical vision-based tracking can provide a high-precision estimation and robustness system in low-textured and low-visibility environments [257].

The existing works utilized ML and AI techniques to extract appearance features, such as color and texture, and combined the extracted features with other semantic information for tracking [192]. Study in [119] developed a deep learning framework based on the cross-frame keypoint-based detection network and spatial motion information-guided tracking network. The developed framework considers the vehicles' driving status and appearances extracted from satellite videos. Yang et al. [447] improved a collaborative sensing system that integrates a customized metric-learning vision-based vehicle re-identification method to extract vehicle features. The system combines vehicle appearance with traffic network connection as features, which achieves high tracking accuracy.

To the best of our knowledge, no existing LPPM can prevent adversaries from tracking vehicles based on their appearance features.

2.1.1.3 Vehicle Sensor-Based Tracking

On-Board Diagnostic (OBD) readers can be utilized to obtain the data of Electronic Control Units (ECU) in the Controller Area Network (CAN) [435]. The easiest way to locate vehicles is by reading data from the GPS devices, but vehicle manufacturers have divided CAN into sub-networks to harden the vehicles against single sensor tracking [308, 210]. Nevertheless, multiple OBD readers (e.g., inertial, heading, pressure, and speed sensors) can be combined to realize localization [108, 230].

It has been demonstrated in the literature that in-vehicle communication systems such as CAN, entertainment applications, and drivers' mobile phones can be used to locate vehicles [71, 275, 337, 425, 298, 151, 227, 140]. For example, the Tegaron from Daimler Chrysler has navigation infrastructures and can exchange data with Tegaron's control centre [71]. In this way, vehicle localization can be achieved either by the in-vehicle communication systems from the third party or by the mobile communication linkage [275]. The Tire Pressure Monitor System (TPMS), as part of in-vehicle wireless networks, is another potential attack surface to track vehicle locations [337]. The identifiers of the TPMS sensors are unique, and the protocol is vulnerable to reverse engineering, so the Road-Side Unite (RSU) can launch a passive tracking system to capture TPMS packets and map a route with the unique sensor identifiers [425, 298].

Mobile phones with GPS play an important role in the side channel attack to infer the driving trajectory. The adversaries can use the accelerometer data from the mobile [151] or the angle matching of the route with the mobile magnetometer sensor data [227] to track a car. Guha et al. [140] leveraged the data from the accelerometer and gyroscope sensors to detect vehicular movements, stops and turns to match the driving path. Besides the existing vehicle onboard hardware, vehicular services and applications can also provide information to track vehicle trajectories during side-channel attacks. The in-vehicle auxiliary applications are required to support the autonomous driving functions by delivering services in the passengers' services, vehicles' services, intelligent communication options, and intelligent resource allocation [141]. However, these in-vehicle auxiliary applications will generate more data embedding the vehicle state and location, which increases the privacy breach risk.

To the best of our knowledge, there are no universal solutions defending against location privacy attacks in the hardware-based physical layer. The existing in-vehicle intrusion detection systems, such as gateway-, ECU-, and CAN-based [243], focus on detecting malicious data transmitted in-vehicle and can hardly prevent the above passive side-channel attacks on location privacy.

2.1.1.4 Cellular Radio-Based Tracking

Antenna systems are designed to track and steer signals from vehicles [187]. For tracking systems and localization purposes, it is most desirable to combine a small compact omnidirectional sensor array using beamforming techniques [14]. However, the omnidirectional antenna is susceptible to signal loss in long-distance communication [187]. In the case of long-distance communication, the existing studies utilized a directional high gain antenna that focuses signal energy in a particular direction [314]. Burghal et al. [65] realized a relative vehicular localization using the channel state information from multiple-antenna transceivers. The authors used feed-forward neural networks to reduce the number of trainable parameters. The mmWave communication can also contribute to vehicle localization, as electrically steerable directivity of phased arrays offers a direction finding [112]. Several direction findings of base stations can be merged in a cross-bearing manner to localize the vehicles [159].

To the best of our knowledge, no existing studies have considered the adversaries with cellular radio-based localization.

2.1.1.5 Upper-Layer Message-Based Tracking

The upper-layer message-based tracking techniques allow the adversaries to obtain vehicles' location data by eavesdropping on the V2X communications or attacking the entities. In this subsection, we discuss upper-layer message-based tracking techniques by analyzing the adversaries mentioned in most existing works, as shown in Figure 2.1. The adversaries are

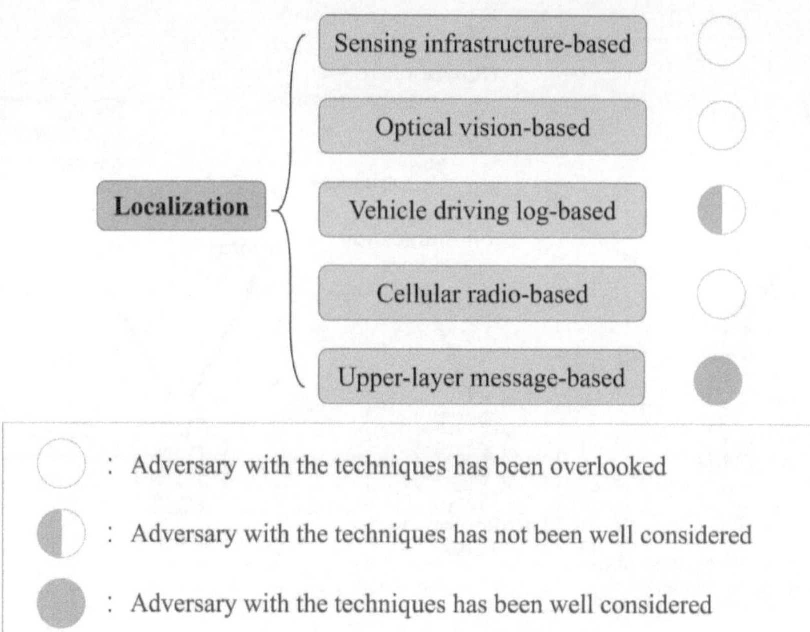

FIGURE 2.1: Localization techniques.

classified based on the behaviors and the size of the monitor region.

An LBS system is composed of three main components: drivers, servers, and networks, as depicted in Figure 2.2. The positioning system plays a crucial role by determining the geographic locations of users and relaying this information to them, with users being represented by mobile devices, wearable devices, and vehicles. The users connect to the communication network, which includes Wi-Fi, the Internet, and cellular networks, enabling the necessary data exchange. The content/data provider is essential in this system, supplying critical information such as maps, traffic data, and points of interest to the LBS server. The LBS server then processes this information, tailoring and delivering services to the users based on their locations. To protect user privacy, the location privacy server acts as an intermediary, ensuring that location data is safeguarded before it reaches the LBS server. However, each component of the LBS system is susceptible to threats from adversaries. For example, eavesdropping on the communication network can lead to the interception of messages transmitted between nodes, potentially exposing sensitive information. Similarly, compromising the positioning system or the LBS server can result in inaccurate location data or unauthorized access to user information.

Based on the behaviors, the adversaries can be classified into active attackers and passive attackers [40]. The active adversaries focus on disrupting network communication, falsifying communication data, and inserting fake messages by capturing or cloning the legal drivers [176]. Compared with the active adversaries, the passive adversaries aim to monitor and analyze the data traffic to discover drivers' positions by eavesdropping [340]. The passive adversaries do not disrupt or interfere with communication directly [169]. The majority of existing LPPMs focus more on the passive adversaries than the active ones.

According to the size of the region that adversaries can monitor, passive adversaries can be classified as Global Passive Adversaries (GPA) and Local Passive Adversaries (LPA) [59], as illustrated in Figure 2.3. The figure delineates the roles and interactions between various entities involved in vehicle location privacy. The GPA, depicted as an entity with extensive monitoring capabilities, can eavesdrop on data transmitted across entire networks,

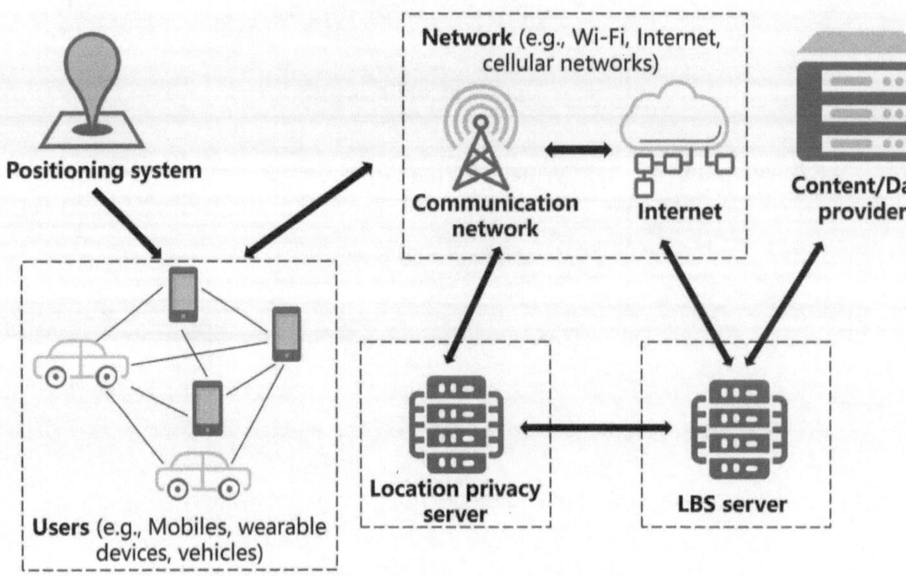

FIGURE 2.2: System model of LBS.

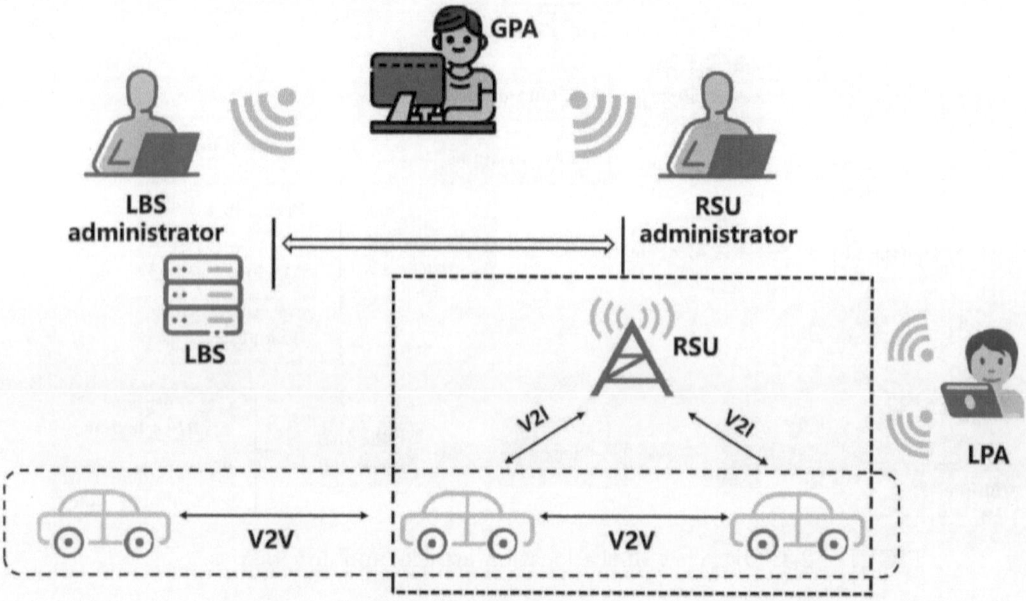

FIGURE 2.3: GPA and LPA for vehicle location privacy.

possessing full knowledge of the road infrastructure by monitoring data, obtaining legitimate authorities, or hacking into applications. This extensive monitoring allows the GPA to gather information over extended periods, such as hours, months, or even years [250]. Due to its extensive reach and capabilities, the GPA model is often considered the strongest adversary in many privacy methods [56, 417, 57]. However, the GPA scenario is theoretical mainly because implementing such a comprehensive surveillance system requires deploying a vast amount of equipment [340]. In contrast, the LPA is represented in the figure as an entity with a much more confined scope, capable of eavesdropping only within a relatively small communication range using limited equipment. The LPA typically monitors specific areas or segments of the network, making it a more practical but less powerful adversary compared to the GPA. The figure also highlights the interactions between different administrators and systems, such as the LBS administrator, RSU administrator, V2V, and V2I.

Figure 2.4 shows an overview of the four main steps of location attacks. The adversaries obtain the location information of the drivers by collecting, eavesdropping, and compromising the vehicle's information. By analyzing the obtained information, the adversaries can gain useful knowledge, with which the adversaries can use methods, e.g., context linking, probability theory, machine learning, and (fake) peer users, to infer the location information of the driver. And then, the identities, discretized trajectory points, and continuous trajectories would be exposed to the adversaries.

2.1.2 Location Privacy Threat in Vehicular Networks

The existing vehicular networks are under various location privacy attacks. One of the most straightforward vehicle tracking methods is illegally setting a Global Navigation Satellite System (GNSS) on it [64]. The adversaries can obtain the drivers' real-time locations by this method. However, this method is infeasible because the adversaries have to set GNSS equipment for each target vehicle.

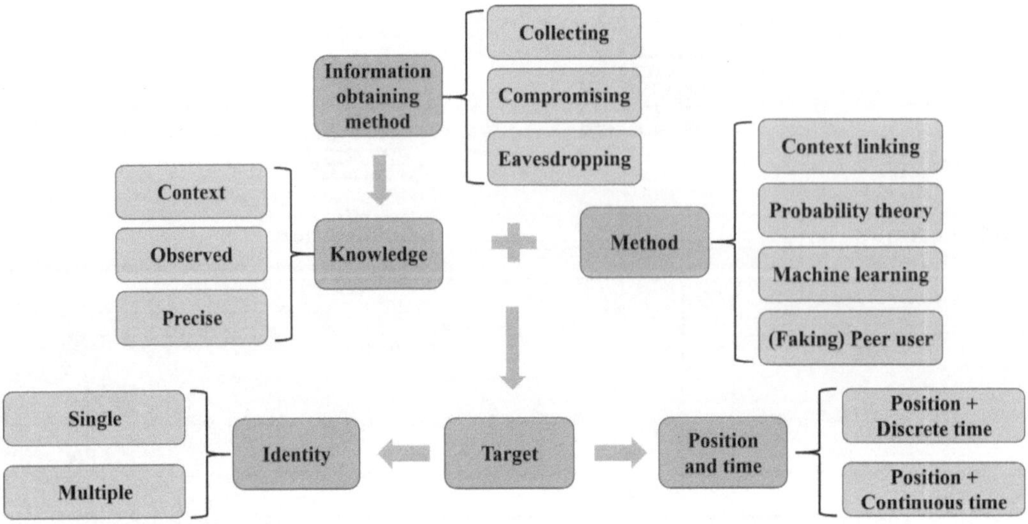

FIGURE 2.4: Overview of the location attacks and adversaries.

2.1.2.1 Passive and Active Attack

The location privacy attacks can be viewed as a typical Multiple Target Tracking (MTT) issue, which assumes a set of noisy measurements or observations detected periodically by a sensor [111]. The adversaries obtain the best estimation of the driver's state and association probability through various location privacy attacks. The location privacy attacks can be classified into passive attacks and active attacks [176], as shown in Figure 2.5.

The adversaries with passive attacks aim to monitor and analyze the data traffic to discover the drivers' positions by eavesdropping that do not disrupt or interfere with communication directly [340]. The passive attacks in vehicular networks can be classified as follows [176].

• *Wireless Eavesdropping Attack:* The adversaries can easily eavesdrop on vehicular networks due to the open wireless medium. Through eavesdropping, the adversaries can obtain data transmitted in networks for future analysis [300].

• *Tracing Back Attack:* The adversaries can employ a triangulation algorithm to locate the drivers' positions with at least two separate antennas through tracing back attacks [361].

• *Traffic Analysis Attack:* The core nodes in vehicular networks process more traffic flow than ordinary nodes [41]. For example, sources and destinations transmit data at a higher rate speed than ordinary nodes. Thus, the adversaries can discover the targets' locations by analyzing the traffic flow.

• *Packet Analysis Attack:* The adversaries can extract encrypted information from the captured packet, e.g., location information and identification (ID) information [204]. The adversaries can infer the sender's trajectory by analyzing packets' timestamps if the packets have the same ID in two different locations.

• *Back-rolling Attack:* The adversaries can utilize the back-rolling attack if they have more extensive storage than legitimate nodes [176]. The adversaries can record the drivers' historical and new positions to find out the location information of targets.

The active attackers can interfere with vehicular networks (e.g., compromising and cloning legal nodes). The active attacks can be classified as follows [176].

• *Node Compromised Attack:* The active adversaries can capture packets from several legitimate nodes to estimate drivers' locations [465]. The adversaries can also infer the network topology and drivers' locations by interfering with networks through cloning nodes [488].

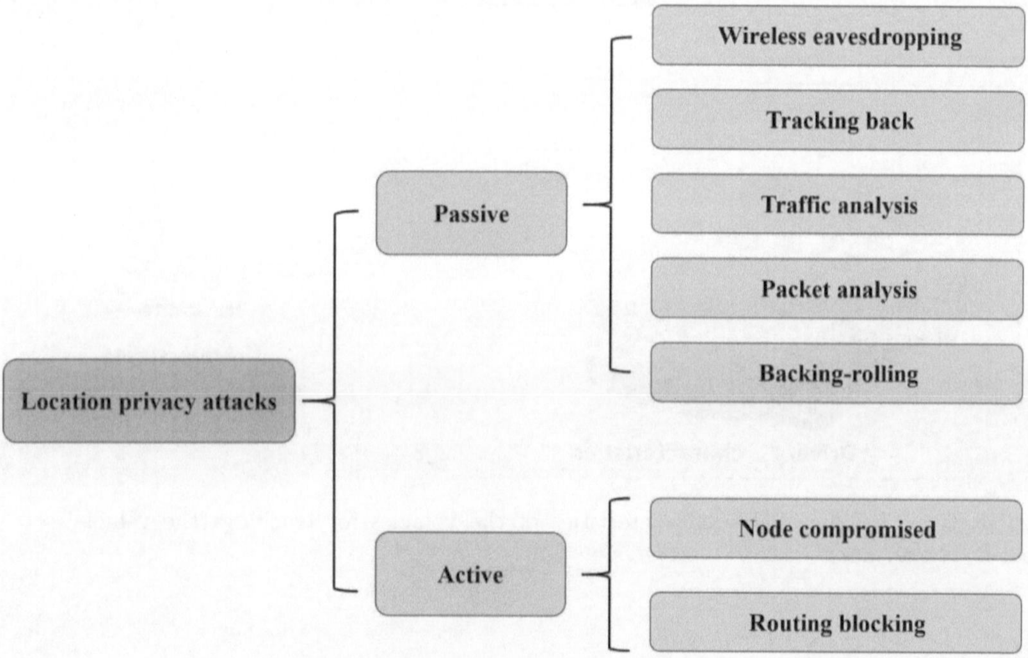

FIGURE 2.5: Major privacy attacks.

- *Routing Blocking Attack:* The routing-blocking attacks track back the target drivers hop by hop to obtain the drivers' data. The adversaries can trigger a fixed routing block and monitor the data through routing blocking attacks [75].

2.1.2.2 Attack Based on Physical Characteristic

The physical characteristics can disclose the drivers' location information. As shown in Figure 2.6,[1] the adversaries can track the drivers' locations by observing vehicles' shapes, license plates, and other characteristics. The tracking technologies (e.g., Automatic Number Plate Recognition (ANPR)) make it possible to track the vehicles with their physical characteristics [494]. The adversaries can estimate the vehicle's trajectory when the vehicle is identified at several different locations. The adversaries' estimation error decreases with the number of exposed locations increases. The obtained data can be utilized for later tracking attacks even if a vehicle is identified at one location [317].

2.1.2.3 Attack on Inside-Vehicle Message

Inside-vehicle communication messages indicate the drivers' location data since the messages represent vehicle speed and steering angle [238]. The adversaries can obtain the data in Electronic Control Units (ECU) by illegal OBD reader on the OBD port or through vulnerabilities [287]. Inside sensors can transmit data in wireless networks [64]. For example, the tyre pressure monitoring system can broadcast unencrypted data within 40 meters. Remote keyless entry technology sends unique identifiers to unlock vehicles through short-range broadcasts. The unique identifiers can be used to track the vehicles [105].

[1]The vehicle in the figure is owned by the authors.

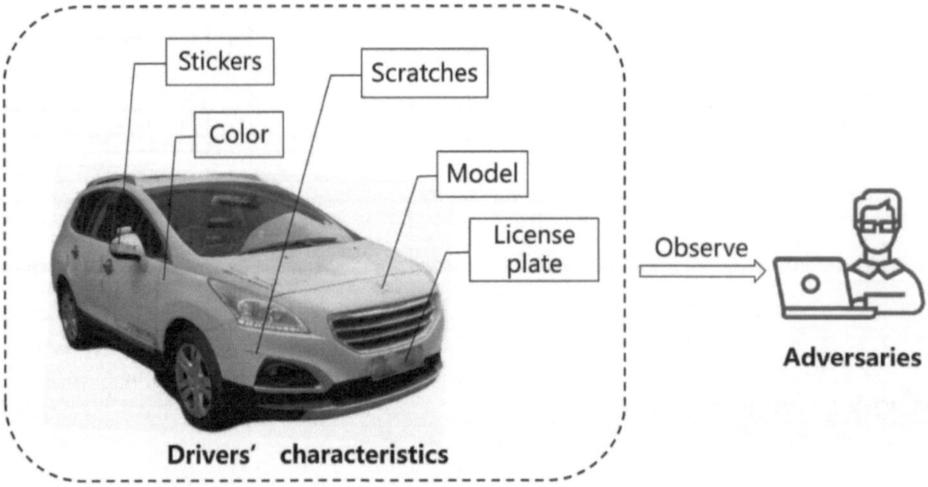

FIGURE 2.6: The identifiable characteristics of the vehicles for tracking (the vehicle is owned by the author).

2.1.2.4 Attack on V2X Message

V2X communications in vehicular networks are at the risk of disclosing location privacy, as shown in Figure 2.7. Cooperative Awareness Messages (CAM) and Decentralized Environmental Notification Messages (DENM) are periodically broadcast by intelligent transport system stations [277]. CAM and DENM messages consist of the vehicles' status information, such as time, location, and speed. CAM and DENM are necessary for the security services in vehicular networks, e.g., road condition warning and hazardous location warning [6]. CAM and DENM, which are unencrypted to decrease time delay, contain the drivers' digital signatures for message verification [244]. The digital signatures of CAM and DENM disclose the vehicles' accurate location data through high-time resolution [64]. The Medium Access Control (MAC) layer protocols can disclose drivers' locations, such as the Time-Division Multiple Access (TDMA) [297]. TDMA MAC slot is broadcast as the vehicles' identifications, which makes it easy for the adversaries to track the vehicles' trajectories by eavesdropping on the wireless channels [20].

The additional devices and peripherals inside the vehicles can also disclose the drivers' location privacy, as shown in Figure 2.7. The adversaries or telecommunication companies can recognize the drivers' regions when they use their mobile phones within the region of a signal tower [339]. And then, the adversaries can achieve high-precision localization by using multiple signal towers. Another unique identification that can be utilized for tracking is International Mobile Subscriber Identity (IMSI) [161]. IMSI can be captured by multiple IMSI catchers, which lead to vehicle tracking if the drivers take their mobile phones into their vehicles [246]. Bluetooth sensors in mobile phones also periodically broadcast advertisement packets containing the devices' MAC addresses [481]. The adversaries can calculate the drivers' trajectories if they can record the Bluetooth MAC addresses of the drivers [443].

The adversaries can track vehicles without the content of the transmitted data. As shown in Figures 2.5 and 2.6, the signal direction context is sufficient for the adversaries to realize traceback attacks within the wireless sensor networks [406]. In vehicular networks, the adversaries can also infer the position changes and velocity changes with received CAM.

2.1.2.5 Security Attacks in Vehicular Networks

In the intricate ecosystem of vehicular networks, security attacks manifest through a spectrum of tactics, ranging from data breaches to unauthorized remote control of vehicles and

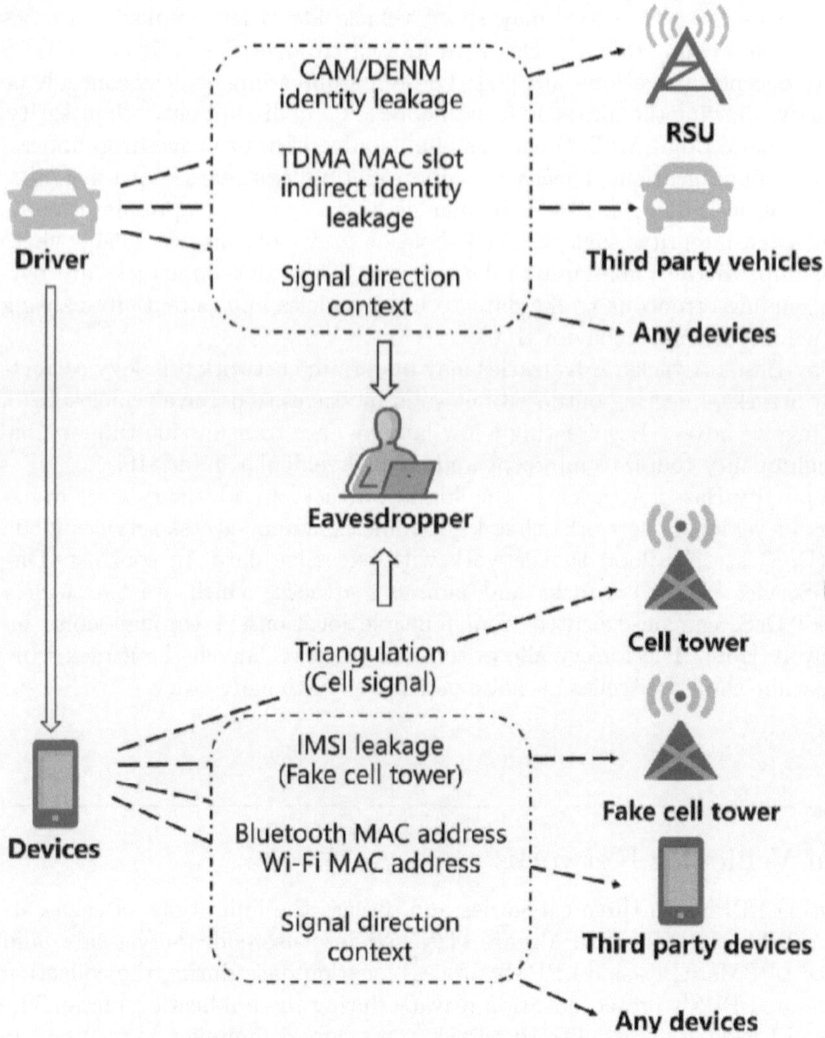

FIGURE 2.7: V2X and in-vehicle communication threats.

severely compromising network integrity and vehicular operation. The attacks can lead to serious privacy and security issues, including unauthorized access and malicious activities aimed at both data and physical control of vehicles. These attacks can override or bypass a driver's decisions, leading to potential adverse remote control or manipulation of vehicle functions. In platoon scenarios, cybersecurity breaches could result in a loss of formation control, unauthorized lane changes, or platoon disbandment—each with potentially catastrophic outcomes. Real-world instances, like certain Sydney roads granting green lights to approaching trucks to maintain traffic fluidity, highlight the potential impact of such attacks on city logistics [2]. Beyond general data privacy, individual vehicle tracking poses direct threats to driver safety and privacy, with identity-based and message replay attacks being prime examples of such vulnerabilities in vehicular communications.

Security measures adopted in vehicular networks strive to safeguard data against unauthorized access, manipulation, or disclosure, fostering trust within the network [118]. The array of security attacks, detailed in [145, 23] are multifaceted.

[2] `https://www.governmentnews.com.au/new-technology-sees-trucks-talk-to-traffic-lights/`

• Identity-based Attacks: Adversaries may spoof vehicle identities, employing tactics like Sybil attacks to create counterfeit vehicles with fabricated identities [307] or use GPS spoofing to generate deceptive location data [27]. These masquerading attacks conceal the attacker's true identity, allowing them to send fraudulent data and disrupt network integrity. Additionally, Man-in-the-Middle (MITM) attacks enable adversaries to eavesdrop and respond maliciously in communications, effectively impersonating legitimate vehicles [173].

• Message Modification-Based Attacks: Vehicular networks can be compromised by attacks that alter message integrity, such as black hole or gray hole attacks [364], where adversaries drop or unpredictably manipulate data routing [272]. Illusion attacks are particularly insidious, sending erroneous traffic data to legal vehicles and potentially causing accidents by influencing vehicular behavior [190].

• Message Replay-Based Attacks: Adversaries may undermine network topology correctness through replay attacks, injecting outdated but valid messages to deceive vehicles [329]. Wormhole attacks involve adversaries creating a low-latency data transmission tunnel, misrepresented as a high-quality route, to intercept and control vehicular data [310].

• Network Availability-Based Attacks: In this kind of attack, an adversary aims to exhaust the resources of vehicular networks that legal vehicles cannot access services [260]. Denial-of-Service (DoS) attacks flood the network with excessive data. In contrast, Distributed Denial-of-Service (DDoS) attacks and jamming attacks, which are two widely observed variants of DoS, can be executed from multiple locations or through noise interference, respectively [260]. The former allows the adversary to launch the attack from different locations, while the latter relies on noise generated by the adversary.

2.2 LPPM in Vehicular Networks

We assess the existing LPPMs in three categories and discuss the limitations of each category, as shown in Table 2.2. The LPPMs are classified into user-side, server-side, and user-server-interface LPPMs. User-side LPPMs process location data during the collection phase, while server-side LPPMs protect location privacy during the publication phase. The user-server-interface LPPMs use trusted third parties and secure communications to realize location privacy protection. We summarize the limitations of LPPMs from the aspect of localization and communication requirements, and we review the method to balance location privacy and data utility from the aspects of theory and practice. The theoretical methods are classified into blockchain, adaptive parameters, hybrid, encryption, element simplification, and virtual nodes, where we provide existing works to illustrate the method in detail. The real-world in-vehicle tracking techniques, i.e., COVID-19 tracking applications, are used as examples to discuss the balance between location privacy and quality of service in practice.

2.2.1 User-Side LPPMs

The user-side LPPMs aim to protect location privacy on the driver side in the collection phase. The popular user-side LPPMs include pass-and-run, certificate, secure computation, and data perturbation.

2.2.1.1 Pass-and-Run

Pass-and-run aims to transmit the information through other vehicles instead of sending data directly to LBSs [101]. The pass-and-run is first proposed in [100]. The authors treat vehicular networks as delay-tolerant networks. A vehicle can decide whether to pass the message to other vehicles or submit the message to LBSs. Lu et al. [248] proposed a lightweight pass-and-run method where the location data is perturbed according to the mobility of vehicles and the delay of transmission. The vehicles have two strategies in the method, i.e., a greedy strategy and a random strategy, which are selected based on the drivers' requirements. The authors utilize an asymmetric encryption algorithm to protect the passed message. Nevertheless, the transmission delay of the method is high.

TABLE 2.2: An overview of the existing LPPMs.

Category	**Technique**	**Demand**	**Benefits**	**Limitations**
User-side LPPMs	Pass-and-Run [100, 248]	Transmit location data through other vehicles	Transmission delay breaks the linkage of location information	High communication complexity
	Certificates for Privacy [38, 147, 180, 181, 261, 267, 280, 281, 299, 353, 382, 432, 484]	Authentication	Provide both location privacy and authentication	High computational and storage consumption
	Secure Computation [489]	Operate encrypted data directly	Operation is flexible and does not reveal private data	High computational and transmission delay
	Data perturbation [290, 114, 32, 235, 490, 218, 222, 418, 223]	Apply LBSs with fake location data rather than the actual version	Considers the prior knowledge of the adversaries and does not need a trusted third party	Low data utility

continued on next page

continued from previous page

Server-side LPPMs	Statistical Disclosure Control [214, 251, 29, 216, 270, 492, 368, 134, 274, 400]	Anonymize or obfuscates data in the dataset	Keep the general statistical features of the dataset	Need a trusted third party
	Homomorphic Encryption [464, 380, 457, 311, 276, 116]	Operate encrypted data directly on the server side	The operation could be flexible and does not need to decrypt data	High computational consumption and transmission delay
	Private Information Retrieval [392]	Hide the requested items	The drivers can apply for LBSs without disclosing their requirements	High computational consumption
	Searchable Encryption [77]	Hide plaintext keywords in searching	Can be combined with other methods	Low accuracy
User-server-interface LPPMs	Secure Communication	Use traditional protocols or end-to-end	Techniques have been developed in past decades encrypted services	There are some limitations of the traditional protocols

continued on next page

continued from previous page				
	Trusted Third Party	Introduce trusted third parties to assist communication	High efficiency	It is an ideal environment

2.2.1.2 Certificates for Privacy

Certificates, also known as privacy-preserving attribute-based credentials, are cryptographic mechanisms [319] that allow the drivers to obtain certified credentials for their attributes from trusted issuers, only reveal required information satisfying the requested LBSs' predicates [395]. The design of certificates relies on the use of malleable signatures [205], as follows.

• *Attribute-Based Signature:* Shahandashti et al. [353] introduce the concept of attribute-based signature, which allows the drivers to sign messages with fine-grained control over identifying information [261]. The attribute-based signature is calculated based on the attribute value, which can be a binary-bit string [294, 215], or relies on a particular data structure [475]. Maji et al. [261] and El Kaafarani et al. [180] introduce and formalize the anonymity of the attribute-based signature that the signature should not reveal the driver's identity or the used attributes. The authors point out that the adversaries can track the drivers' locations by identifying their actual attributes. Considering the same privacy issues, Kaaniche et al. [181] improve the anonymous attribute-based signature by using a concrete mathematical construction based on standard assumptions and the random oracle model. The improved model significantly improves the location privacy and identity privacy of the attribute-based signature.

• *Group Signature:* Group signature allows a driver of a group to anonymously sign messages on behalf of the group [44]. Then, any verified vehicle can confirm that the signature is generated by a legal group member without requiring the signature to be identified by the signer. Zheng et al. [484] improve a linkable group signature that achieves anonymity, auditing, and tracing functions for the communication sender. The improved framework increases communication efficiency and security by employing cryptography modules of blind signature, public-key encryption, trapdoor indicative commitment, and signature of knowledge. Hakeem et al. [147] employ a short-size signature to broadcast authentication information over multiple zones. The authors decrease the signature generation time and verification time by utilizing bilinear pair cryptography. Wu et al. [432] point out that bilinear pair cryptography is complex for OBUs and RSUs. The authors use an elliptic curve cryptosystem in the authentication process that decreases the computational complexity of group signatures. Mundhe et al. [280] develop a ring signature that provides unconditional privacy to the drivers by transmitting the message through verified legal vehicles. The authors also utilize pseudonyms to increase the privacy-preserving capability of the ring signature. Mundhe et al. [281] enable RSUs to participate in the signature generation process to decrease the authentication delay of the ring-signature-based LPPMs. Mei et al. [267] adopt the full aggregation certificateless signature technology to reduce the bandwidth resources overhead in the transmission of the certificate.

• *Sanitizable Signature:* Ateniese et al. [38] were the first to develop sanitizable signatures in 2005. In the sanitizable signature, authorized semi-trusted censors can modify part of a signed message in a limited and controlled fashion without interacting with the original signer. Pamies et al. [299] combine the log anonymity and sterilizable signature to protect sensitive data, e.g., identity and location. The authors hide authenticated identification when transferring the data streams from the local node to remote storage servers.

• *Blind Signature:* For the blind signature, the content of a message is disguised before it is signed, which means the signer and message author are different parties [412]. Sun et al. [382] improve a fog-computing-based crowdsensing architecture where identity and location privacy are protected by partially blind signature authentication. The authors use zero-knowledge verification to improve the security of the blind signature.

In summary, the limitation of the signature-based methods is that certificate management leads to high computational and storage consumption.

2.2.1.3 Secure Computation Mechanism

Secure computation techniques protect a driver's location privacy by processing the location data, which was first introduced and formalized in 1982 based on the millionaire problem [449]. Zhuo et al. [489] improve a multi-key secure outsourced computation scheme that does not need the interaction between the LBS servers and drivers. The improved scheme avoids duplicating and useless encrypted LBS messages before verifying identities, which ensures the quality of LBSs.

2.2.1.4 Data Perturbation

The obfuscation-based LPPMs allow the drivers to perturb their location before sending the location information to LBS servers [255]. The data perturbation can be utilized on either the user side or the server side. In user-side data perturbation, drivers do not need to trust any external entity by sending obfuscated location data to LBS servers [197]. The data perturbations can ensure personalized location privacy, which offers location privacy protection with acceptable data utility [255]. The user-side data perturbation is summarized as follows.

• *Dummy-Based Method:* Dummy-based methods do not need a trusted third party or key sharing step [431]. Niu et al. [290] improve a dummy-location selection algorithm to achieve k-anonymity for the drivers. The algorithm selects dummy locations according to the entropy of anonymity. The limitation of the dummy-based methods is the insecurity of submitting spatio-temporal correlation in sequential LBS requests. Liu et al. [235] point out that time reachability, direction similarity, and the degree should be a concern when filtering the dummies to solve this limitation. After the filtering strategy, the candidate dummies can provide high-privacy protection for the driver's location information. The storage cost of the filtering strategy is acceptable, but the computation delay could be high.

• *Local Differential Privacy (LDP):* Differential privacy can bind the knowledge obtained by the adversaries but could decrease the quality of LBSs [433]. From the LBS providers' perspective, applications can provide a high quality of services if high accuracy is received, but most LBSs can accept location data that is not entirely accurate [139]. The differential privacy can be defined as follows,

Definition 2.1 (Differential Privacy (DP)) *A mechanism $\mathcal{M}$ satisfies ϵ-DP if and only if, for any pair of data x_i and x_j, we have*

$$\frac{\Pr[\mathcal{M}(x_i) \to y]}{\Pr[\mathcal{M}(x_j) \to y]} \leq e^{\epsilon}, \tag{2.1}$$

where y is the output of the mechanism.

LDP is a distributed variant of traditional differential privacy that allows the drivers to perturb their location information before sending it to servers [98]. Erlingsson et al. [114] amplify the privacy to achieve high privacy-preserving capability from local differential privacy by combining differential privacy and anonymity. The authors point out that location privacy security can be achieved without adding any significant noise if the method employs LDP on the client side and a shuffling strategy on the server side. Nevertheless, the authors ignore the spatio-temporal correlation in the road network.

• *Geo-Indistinguishability (Geo-I):* Geo-I is first proposed by Andrés [32]. Geo-I allows the drivers to enjoy ϵr-differential privacy in the given obfuscate radius r with a privacy budget ϵ, which is as given by

Definition 2.2 (Geo-I) *A mechanism $\mathcal{M}$ satisfies ϵ-Geo-I if and only if, for any two locations x_i and x_j, the following holds*

$$\frac{\Pr[\mathcal{M}(x_i) \to y]}{\Pr[\mathcal{M}(x_j) \to y]} \le e^{\epsilon d(x_i, x_j)}, \tag{2.2}$$

where y is the output of the mechanism, and $d(x_i, x_j)$ is the distance of x_i and x_j.

Based on the Geo-I, Zhou et al. [490] design a framework to balance the utility and privacy in edge computing. The framework includes privacy-preserving location-based service usage methods and privacy-based service adjustments. The authors add two-dimensional Gaussian noise to shift actual locations. Although the authors consider the trade-off between privacy and service quality, the framework can lead to a high calculation consumption. Based on the background knowledge of trusty serves, Li et al. [217] employ correlation probabilities and correlation transition probabilities to realize Geo-I. The proposed method can provide different privacy-preserving levels for various requirements. The location shift decides the level of privacy protection. The shift is invalid when the driver's location shift is shorter than the threshold. Although the method can provide a different protection level, it ignores road conditions. If the traffic jams and the driver applies LBSs frequently, the obfuscated location will never change by which the adversaries can obtain the driving state. Li et al. [222] improve an enhanced Geo-I definition named Perturbation-Hidden to ensure perturbed locations are valid. The Perturbation-Hidden method transforms the map of road networks into a grid where the acceptable locations are used as the candidate set. Furthermore, dynamic programming is employed to determine the retrieval area to provide accurate LBSs. The authors employ the dynamic programming method to provide the drivers with the shortest radius of retrieval radius. The limitation of the Perturbation-Hidden method is that it may lead to high-privacy costs in privacy-limited regions.

• *Pseudonym:* Pseudonyms are employed as temporary anonymous certificates generated and distributed by the certificate authority [146]. The pseudonym-based LPPMs aim to ensure the unlinkability between the driver's identity and pseudonyms in communication [16]. Wang et al. [418] treated pseudonyms as a long-term identifier to decrease the computation and storage consumption. The authors designed a trigger for pseudonym exchange requests to assist the certificate authority in changing the pseudonym. Vehicles change their pseudonyms when meeting the trigger. The method has a limitation in that a long period of existing pseudonyms may provide a longer track window for the adversaries. Pseudonym-Indistinguishability was first proposed in [223] to ensure strict unlinkability in the pseudonym swap process. The pseudonym swap process satisfies differential privacy. The adversaries cannot link the pseudonyms after swapping, even if the driving states of the two vehicles are similar. The Pseudonym-Indistinguishability method can provide a high swap complete probability with fewer pseudonyms. The limitation of the method is that the authors ignore the conflict in the pseudonym swap. There are two weaknesses of the pseudonym-based LPPMs. One weakness is that the pseudonym-based LPPMs must manage the vehicles' pseudonyms, leading to high computation and storage consumption [90]. Another weakness is that the pseudonym-based LPPMs cannot ensure unlinkability in the tracking attack [256].

2.2.2 Server-Side LPPMs

By using server-side LPPMs, service providers are required to perform additional processing on their clients' hosted data [213]. Service providers can achieve this by anonymizing databases, removing identifying traces or encrypting data contents.

2.2.2.1 Statistical Disclosure Control (SDC)

SDC mechanisms are mainly used to protect data within statistical databases, which balances data utility and drivers' location privacy [175]. Generally, the output of the SDC mechanisms ensures that the databases do not reveal information related to a specific driver [70]. Database anonymity and differential privacy are the two most popular techniques in the SDC mechanism, as follows.

• *k-anonymity:* k-anonymity LPPMs can provide personalized and accurate query results without key sharing [473]. k-anonymity LPPMs protect driver's location data in k vehicles to increase the estimation error of the adversaries [436]. The k vehicles are selected according to the closest historical request probability based on the maximum entropy principle [5]. The existing k-anonymity methods can be classified into centralized k-anonymity and distributed k-anonymity [373]. The centralized k-anonymity methods need anonymous cloaking servers, while the distributed k-anonymity methods ask the participants to unite [34]. Li et al. [214] propose a distributed architecture utilizing the blockchain. The proposed architecture records the hashed safety beacon messages to ensure integrity by reducing storage consumption and processing time. The authors employ k-anonymity to gather and upload safety beacon messages of a group of vehicles. The proposed architecture does not need a trusted third party. Luo et al. [251] improve a blockchain-enabled k-anonymity method to ensure trustworthiness. The proposed method calculates the trust level of vehicles based on the historical trust data and the trust degree reflecting factors. The authors establish a blockchain structure to record the historical trust information. The authors treat RSUs as distributed k-anonymity databases that provide the recorded data.

• *Mix-zone:* The Mix-zone methods allow a certain number of drivers to change pseudonyms in specific regions [357]. The drivers orderly get into the regions and get out of the regions in a different order [60]. Mix-zone methods cannot against timing and transition attacks that the adversaries can link the driver's pseudonyms at entry and exit points based on the timing knowledge [410, 483]. The adversaries can also employ the continuous query correlation attack to link the old and new pseudonyms [268]. Some pseudonym mappings can be ruled out because of the weighting time, traffic conditions, and time constraints [91]. Amro [29] proposed a Mix-zone method that introduces fixed transceivers to decrease the influence of traffic conditions. Legitimate virtual vehicles join in the pseudonym swap process when there is a lack of physical vehicles. The proposed method requires transceivers that act as physical vehicles. The introduced transceivers use pseudonyms to communicate with RSUs. However, the introduced transceivers bring some security and privacy issues. For example, the pseudonym swap process will be insecure if the adversaries obtain the pseudonyms of the transceivers. A group-based dynamic Mix-zone method is proposed to protect location privacy in resource-limited regions [216]. The proposed method allows vehicles to transmit encrypted information in the silent region. The method is personalized and considers the expiration time of the drivers' pseudonyms.

• *Other Anonymity-Base LPPM:* Meng et al. [270] proposed a method to protect the driver's location privacy in navigation services. The method extends anonymous authentication that supports a request-limiting property. The method protects both the location and route information of the users. The method can effectively protect identity privacy, location privacy, route privacy, unlinkability, and confidentiality. Zhu et al. [492] proposed an Anonymous Smart-parking And Payment (ASAP) method to realize smart parking navigation. The ASAP method applies servers anonymously that the drivers transmit their locations into a region with the cloaking method to hash and encrypt the location data. Locations and routes can be explored under similar continuous queries, although the actual location is cloaked. Singh et al. [368] proposed a Masqueraded Probabilistic Flooding for Source-Location Privacy (MPFSLP) method to reduce communication costs. The method can ensure non-repudiation, message authentication, integrity, and non-traceability. In the MPFSLP method, re-sending is defined as replacing the forwarding process. The authors point out that the vehicles do not require a relation between the location data and the identities of other vehicles. The MPFSLP method allows each node to send packets by masquerading as probabilistic flooding instead of generating fake packets. The vehicles can re-send previous messages to complicate the identity.

• *General Differential Privacy:* Differential privacy can also be used on the server side. Soheila et al. [134] proposed a Differentially Private Data Streaming (DPDS) system to aggregate the data in vehicular networks. By considering the data correlation, vehicles in the DPDS system set up several groups in which the group leaders send members' compressed information. The compressed information reduces the noise value, but its limitation is that the compressed data size is similar to the original data. In [274], the authors proposed a protection framework to protect privacy in edge computing. They also proposed a data transmission method based on the noise quadtree and the Hilbert curve. The proposed method can improve the efficiency of location data publishing. Nonetheless, they only consider the two-dimensional space, which may not satisfy the real-world environment.

• *Joint Differential Privacy:* Joint differential privacy limits the single driver's manipulation power and reduces the impact of the single driver's false report [152]. Based on the joint differential privacy, a study in [400] proposed a scheduling protocol to protect location privacy and minimize vehicle miles in the ridesharing services. The proposed method employs private dual decomposition technology and driver clustering to improve the scheduling performance. They also design the private ride assignment and driver grouping methods to improve privacy-preserving capability.

2.2.2.2 Homomorphic Encryption

Encryption is widely used to protect the location privacy of vehicles in data dissemination and computation, but data decryption needs to be more flexible and secure [384]. Homomorphic Encryption (HE) schemes can directly analyze ciphertexts by mirroring the corresponding operations on the plaintexts, which means the decrypted values of computation results on ciphertexts of location data in the HE schemes correspond to the values of operations on the plaintexts [429]. The most common definition of the HE is given by

Definition 2.3 *Let $\mathcal{P}$ and C be a set of plaintexts and ciphertexts, respectively. An encryption mechanism $\mathcal{M}$ satisfies homomorphic if and only if, for any given encryption key k, and any pair of data x_i and x_j, the following holds: $\forall x_i, x_j$*

$$\mathcal{M}(x_i \odot_{\mathcal{P}} x_j) \leftarrow \mathcal{M}(x_i) \odot_{\mathrm{C}} \mathcal{M}(x_j), \tag{2.3}$$

for some operators $\odot_{\mathcal{P}}$ in $\mathcal{P}$ and $\odot_{\mathrm{C}}$ in C, where $\rightarrow$ means that the left-hand side can be directly computed from the right-hand side.

The HE schemes can be classified into fully homomorphic encryption and partially homomorphic encryption [258], as follows.

• *Partial Homomorphic Encryption (PHE):* PHE schemes are the earliest type of HE schemes and only support homomorphic addition or homomorphic multiplication [384]. Yucel et al. [464] use PHE to solve the location privacy issues in the charging station scheduling, where frequent scheduling could expose the drivers' location data. The authors first hide the drivers' location information and then process the hidden information with PHE. The proposed scheme allows the drivers to join and leave the charging station dynamically.

• *Somewhat Homomorphic Encryption (SHE):* SHE schemes support a finite number of homomorphic addition and homomorphic multiplication [384]. The work in [380] provides an integer-based SHE scheme based on the duration between each data transmission from the sensor and the data packaging method. By using the proposed scheme, the authors improve the efficiency of the SHE algorithm by reducing the encryption consumption of the sensor data. Yu et al. [457] focus on the scenario of an online ride-hailing service that allows a group of passengers to share a vehicle with a minimum aggregate distance. The authors use the SHE schemes with ciphertext packing in such a scenario to calculate encrypted aggregate distance. The ciphertext packing is utilized to ensure that the actual value of the aggregate distance cannot be leaked.

• *Fully Homomorphic Encryption (FHE):* FHE schemes support an infinite number of homomorphic [384], but it has limitations of low efficiency and high computational consumption [144]. To decrease the communication overhead of the FHE scheme, Perma et al. [311] combine FHE with pseudonyms. The proposed method uses FHE to encrypt location data transmitted in vehicular networks, which also avoids the linkage between different pseudonyms of the same vehicle. By considering that all LBSs have a deadline and mobility constraints, Mohammed et al. [276] proposed a cost-efficient vehicular fog cloud computing method where the FHE is utilized for security. Farouk et al. [116] utilize a network simulator to simulate urban mobility, and a cloud simulation to simulate the real world and protect LBS queries with FHE based on the advanced encryption standard. The authors outsource encrypted location data to the cloud server so the drivers can securely obtain accurate LBS query results.

Nevertheless, there is no efficient homomorphic encryption-based secure computation method in road networks [384].

2.2.2.3 Private Information Retrieval

Private Information Retrieval (PIR) enables the drivers to request data items without revealing which item is retrieved [81, 131, 383]. However, the computational consumption of the PIR could be so high that it is difficult to allocate in practice [392]. To overcome this limitation, Tan et al. [392] focus on road networks where the authors apply the transportation information of road networks as prior knowledge of PIR that significantly decreases the computational cost.

2.2.2.4 Searchable Encryption (SE)

SE schemes are considered as a server-side privacy-enhancing technique, which enables the drivers to keep plaintext keywords safe in searching LBSs [370, 88]. SE schemes can also be combined with other methods or tools to improve precision and efficiency, e.g., group signatures, Cuckoo filter, Pederson commitment, smart contract, public-key encryption, and proxy re-encryption [123, 393]. However, most existing SE schemes would provide wrong results for geometric range searches [424]. To solve this problem, Chen et al. [77] develop a novel SE method under a public-key system supporting arbitrary geometric area searches. The authors avoid false positive results by introducing inner product encryption. The developed method can achieve 100% accuracy in search results.

2.2.3 User-Server-Interface LPPMs

Channel-side techniques encompass mechanisms that act on the security and privacy properties of the set-up communication channel between the server and end-users [396]. Secure communication and trust in the third party are widespread in the user-server-interface LPPMs.

2.2.3.1 Secure Communication

Secure communication can be classified as follows.

• *Encrypted Communication Protocol:* To avoid pervasive communication surveillance, it is essential to emphasize that encrypted channels need to be implemented and configured correctly [293]. In the LBSs of vehicular networks, traditional protocols are widely used [54], e.g., the Transport Layer Security protocol (TLS) [94, 332] and the Secure Shell (SSH) protocols [454, 125]. TLS and SSH rely on public key cryptography techniques so drivers and servers can set up an encrypted channel without sharing secrets [408]. Nevertheless, the traditional protocols still have limitations. For example, SSH requires that the end-users perform periodical verification [199], which is resource-consuming [67].

• *End-to-End Encrypted Services:* End-to-end encrypted services prevent the adversaries from accessing the location data when transferring among end-users, meaning that only the intended receiver can decrypt the data [42]. End-to-end encrypted services are seldom used in vehicular network scenarios currently. However, as future vehicular networks may require vehicles to cooperate with others to complete a task, end-to-end encrypted services could be widely used in the future.

2.2.3.2 Trusted Third Party (TTP)

The TTP techniques, which are commonly based on anonymity and pseudonym, are proposed to prevent malicious entities from obtaining the drivers' sensitive information [430, 313, 182]. A set of malicious entities could be more powerful than the legal drivers in that they could have the capability to trace the drivers and estimate other private information of the drivers [155]. However, the existing studies [236, 49, 325, 172] prefer to protect location privacy without any trusted third party, which is a strict scenario.

2.2.4 Trade-off between Location Privacy and Data Utility

Location privacy and data utility are two competing objectives of vehicular networks. The data utility is one of the significant concerns in location privacy preservation to measure how the privacy-preserving methods influence LBSs. Drivers are increasingly wary of their location privacy, which may discourage their location information from being shared with the LBS platforms. For commercial purposes, LBS platforms desire the actual locations of the drivers. The over-protection will diminish the data utility of the LBSs [85]. Therefore, a privacy-preserving method should maximize data utility with the best privacy-preserving capability.

Many LPPMs focus on the trade-off between location privacy and data utility, as shown in Table 2.3. There are six significant methods to achieve the trade-off between location privacy and data utility.

TABLE 2.3: Existing LPPMs in the trade-off between location privacy and data utility.

Methods	Techniques	LPPM Category	Papers
Blockchain	Distribution Management	k-anonymity	[214, 251, 74]
		others	[249, 58, 262, 342, 412, 393, 228, 411]
Adaptive parameters	Tuning parameters	Pass-and-run	[248]
		Attribute-based signature	[353, 181]
		Geo-I	[490, 217, 222, 255]
		Pseudonym	[223]
		Other anonymity	[492]
		General differential privacy	[134, 274]
		SHE	[380]
		FHE	[276]
Hybrid	Combining multiple methodologies	Group signature	[484, 280]
		Sanitizable signature	[299]
		Dummy-based	[290]
		LDP	[114]
		Joint differential privacy	[400]
		SE	[77]
		Group signature	[432]

continued on next page

continued from previous page			
Encryption optimization	Reducing computational and communication consumption	Blind signature	[382]
		Secure computation	[489]
		Mix-zone	[216]
		PHE and FHE	[464, 311]
Parameter simplification and optimization	Simplifying process	Group signature	[147, 267]
		Blind signature	[382]
		Dummy-based	[235]
		Pseudonym	[418]
		Other anonymity	[368]
		FHE	[116]
		Private information retrieval	[392]
Virtual nodes	Introducing cooperators	Group signature	[281]
		Pseudonym	[418]
		k-anonymity	[29]
		TTP	[49, 172, 236, 325]

2.2.4.1 Blockchain

Blockchain can be used to realize distribution management. LPPMs can access the required information if the information is stored in the blockchain [456]. This technology can decrease time delays and detect malicious vehicles. Luo et al. [251] use blockchain to store the historical trust information of vehicles and employ Dirichlet distribution to allow vehicles to cooperate with others by utilizing anonymity. The authors reduce the communication delay with blockchain-based trust management. Li et al. [214] develop two metrics, i.e., connectivity and average distance, to measure the data utility of k-anonymity. With these two metrics, the authors evaluate the developed framework, which uses blockchain to solve trust management during the exchanging of safety beacon messages. The experimental results show that the blockchain-based k-anonymity framework decreases the data processing time.

Malik et al. [262] avoid the dependency on a trusted third party by developing a blockchain-based authentication and revocation method that decreases computation and communication consumption. Wang et al. [412] combine the blind signature with blockchain to achieve a scheme that has high effectiveness and applicability. The authors use blockchain to store the public key of the vehicle, which can be used for authentication by comparing the calculated Merkel's root value. Tang et al. [393] manipulate group signature and multiple tools when designing a blockchain-based privacy-preserving scope-query searchable encryption scheme, which achieves fairness and accurate parking lot sharing. Chaudhary et al. [74] also use blockchain to overcome the limitations of k-anonymity, while Liang et al. [228] create cloaking region according to the blockchain. Wang et al. [411] evaluate the trustworthiness of vehicles by using blockchain-based RSUs, and Lu et al. [249] use the proofs of presence and absence in blockchain to protect location privacy. The authors ensure the high validity of the two methods and avoid the leakage of identity. Boualouache et al. [58] and Samuel [342] use blockchain to protect location privacy by preventing semantic, linking, and data mining attacks.

2.2.4.2 Adaptive Parameter

Some LPPMs achieve the trade-off between location privacy and data utility by tuning the parameters according to the drivers' requirements. The adaptive parameters-based LPPMs can decrease unnecessary resource consumption and data utility loss. Lu et al. [248] define the quantitative measurements of location privacy and perfect privacy. By using these measurements, vehicles can route location data with the assistance of other vehicles to obtain efficient, secure and accurate LBSs. Shahandashti et al. [353] and Kaaniche et al. [181] permit the vehicles only to reveal required data based on their attributes and requirements, which can achieve high efficiency and data utility without breaching the anonymity of the vehicles. Ma et al. [255], Zhou et al. [490], Li et al. [217], and Li et al. [222] develop the location privacy-preserving mechanisms which can adaptively set the parameters based on the drivers' protection and utility requirements. Li et al. [223] allow the vehicles to swap their pseudonyms according to their driving status, which can achieve high-privacy protection and data utility. Zhu et al. [492] ensure location data utility effectiveness by using a hashmap-based short randomizable signature. By controlling the noise of differential privacy, Ghane et al. [134] and Miao et al. [274] improve the retrieval accuracy and decrease the time complexity to balance the location privacy protection and data utility. Subramaniyaswamy et al. [380] and Mohammed [276] use encrypt location data in the edge of vehicular networks without losing data. The authors reduce the time delay and recourse consumption of encryption.

2.2.4.3 Hybrid Approach

Some LPPMs utilize different methodologies to improve efficiency. Hybrid LPPMs can obtain significant benefits and overcome limitations by mixing various methodologies. Zheng et al. [484] and Mundhe et al. [281] combine group signatures and pseudonyms to achieve identity verification. By using the pseudonym, the signature could be short, and the computational consumption could be reduced. Similarly, the work in [299] jointly uses log anonymization and sanitizable signatures to mitigate privacy threats. Niu et al. [290] realize k-anonymity by utilizing dummy locations, which considers the side information of the vehicles, and Erlingsson et al. [114] use differential privacy to eliminate the signal at the user side. The two methods can prevent adversaries from linking pseudonyms. With joint differential privacy, Tong et al. [400] research on the private spatial index, private distributed optimization, and private dual decomposition techniques to achieve the balance. Chen et al. [77] use SE and computational private information retrieval to decrease computational consumption, ensuring effectiveness, which considers the restriction of road networks as the prior knowledge.

2.2.4.4 Encryption Optimization

The encryption methodologies result in high communication consumption, such as bilinear pair cryptography. Some LPPMs focus on optimizing encryption methods to achieve high data utility and low time delay. Wu et al. [432] avoid the high computational cost of message signing and verification by using the elliptic curve cryptosystem rather than bilinear pairs. Sun et al. [382] introduce encryption-based methods, including zero-knowledge verification, one-way hashing, and homomorphic encryption into the partially blind signature to obtain superior effectiveness of network response and privacy preservation. Zhou et al. [489] employ the non-colluding servers, i.e., the cloud and the cryptographic service provider, to skip the interaction between the vehicle and the LBS servers. The authors eliminate redundant encrypted LBS data before authentication, i.e., ciphertext re-encryption, that achieves practicability and protection. Li et al. [216] suggest the vehicles transmit encrypted data rather than being silent in the mix zone. The authors point out that encryption can decrease the storage consumption of pseudonym management, and increase privacy protection without breaching data utility. Yucel et al. [464] protect location privacy by utilizing homomorphic encryption into dichromatic mutual nearest neighbor assignments, which protect location privacy with low recourse consumption and convergence times. Prema et al. [311] point out that FHE can lead to communication overhead problems. The authors utilize pseudonyms with homomorphic encryption to control the message frequency, which keeps the low overhead of the pseudonym with the high security of the homomorphic encryption.

2.2.4.5 Parameter Simplification and Optimization

The time delay can be significantly decreased when the LPPMs can simplify the process. For example, the non-certificate LPPMs can eliminate the certificate authentication time delay and certificate storage consumption. Hakeem et al. [147] achieve authentication over multiple zones of large-scale BSs by using a single message and short signature with bilinear pairing cryptography and short-size signature. The authors significantly decrease the generation and verification time of the signature. Mei et al. [267] use a certificateless aggregate signature scheme with full aggregation technology to reduce resource consumption. The trade-off between privacy protection and data utility is achieved by considering random oracles under the computational Diffie-Hellman assumption. Sun et al. [382] optimize a fog-bus-based vehicle crowdsensing framework that severs the relationship between the identity and location data. The authors simplify the data process in data reporting, reputation management, and reward issuing so that the effectiveness can be improved.

Liu et al. [235] focus on the time reachability, direction similarity, and in-degree/out-degree of the location data. The developed spatiotemporal correlation-aware LPPM simplifies neighboring location sets to achieve personalized location privacy protection. Wang et al. [418] improve the data utility of pseudonyms by introducing a trigger-based structure, avoiding the frequent pseudonym changing. Singh et al. [368] simplify the required characteristics of the vehicles by using location and speed instead of their identities. The authors allow each vehicle to send data of others masquerading as its own location data, which significantly decreases the traceability of the pseudonym-base LPPM without breaching the data utility. Farouk et al. [116] outsource the location data to a cloud server that prevents the vehicles from sharing their location data with multiple different entities. By only sharing data with the cloud server, the vehicle can obtain LBSs with a low delay. Tan et al. [392]

use the prior knowledge of road networks to decrease the consumption of computational private information retrieval in preprocessing and communication.

2.2.4.6 Virtual Node

Some LPPMs, e.g., k-anonymity, need a specific number of neighbors to provide a high privacy-preserving capability. The LPPMs can generate virtual nodes to act as physical vehicles to overcome this limitation. The virtual nodes occupy limited storage space. Thus, the privacy-preserving capability can be improved without decreasing the data utility. Mundhe et al. [281] avoid the trusted third party in authentication by employing a lattice-based ring signature. The scheme provides a low authentication delay and cost, improving the effectiveness of the data utility for signature-based LPPMs. Wang et al. [418] introduce virtual devices, i.e., triggers, to assist pseudonym changing, which improves the privacy protection of pseudonym-based LPPMs. Zhu et al. [492] distribute fixed mixing zones in road networks to avoid adversaries linking the pseudonyms. The presented method provides stable location privacy preservation, which is not influenced by the number of vehicles in the mixing zone.

In summary, existing methods that balance location privacy and data utility cannot perfectly satisfy the future 5G/6G-enabled vehicular networks. For example, the Geo-I-based LPPMs inject controlled noise in location data when the data is reported to the LBSs. Nevertheless, inaccurate location data will introduce extra data optimization processes, which degrade the consumption of big data analytics in 5G/6G networks [352]. The data utility will be significantly decreased even if the Geo-I-based methods finely tune the adaptive parameters. This is unacceptable to 5G/6G service providers. The virtual nodes-based method needs to provide the expected balance between location privacy-preserving capability and data utility. The reason is that the future 5G/6G-enabled vehicular networks will deploy numerous sensors to monitor the environment so that virtual nodes can be eliminated. Thus, the challenge of balancing location privacy and data utility will become severe as the requirements of high-precision and novel technologies are introduced in future 5G/6G vehicular networks.

2.2.4.7 Example of Localization Application

Different from the theoretical trade-off between location privacy and data utility, practical applications are more interested in data utility. According to [107], it is highlighted that vehicular networks offer potential applications in tracking the spread of diseases, identifying instances of social distancing non-compliance, and facilitating health-related services. This is attributed to individuals' widespread use of public and private transportation in their daily lives. With the sensors equipped on vehicles, the encounter history with social distance can be collected and sent to the authorities [107]. The use of vehicular communication can improve the detection of infection and prevent the spread of the disease. However, tracking also brings challenges in the pandemic scenario, e.g., COVID-19. Location data collection and exchange in the fight against COVID-19 is an excellent example of in-vehicle tracking. Existing applications for epidemic prevention can be classified into centralized and decentralized management. In centralized management cases, the medical information of confirmed cases is monitored by governments or institutions. Governments or institutions will only disclose the trajectory of the patient when a new patient appears. In decentralized management cases, the location data is managed by the drivers. Only if the drivers are infected are their trajectories exposed to authorities. Strict management of such information can save the world from the virus, but the location privacy of the patient is exposed. Table 2.4 shows the benefits and limitations of related applications. In Table 2.4, the existing COVID-19 applications are classified into centralized and decentralized applications.

TABLE 2.4: Trade-off between location privacy and data utility of in-vehicle tracking in practice: COVID-19 applications.

Architecture	**Applications**	**Required Information**	**Limitations**
Centralized	COVIDSafe (Australia) [87]	Encounter information	The personal information of the driver, e.g., identity, phone number, and email. Encounter information that is frequently recorded exposes the driver's trajectories. Private information can be inferred from the reported geographical information. The geographical information is disclosed to the public for epidemic prevention.
	E-Tabib (Azerbaijan) [21]		
	BeAware Bahrain (Bahrain) [265]		
	Corona Tracer BD (Bangladesh) [162]		
	Taiwan Social Distancing (China) [130]		
	TousAntiCovid (France) [348]		
	VirusRadar (Hungary) [160]		
	Rakning C-19 (Iceland) [163]		
	Smittestopp (Norway) [231]		
	BlueTrace (Singapore) [48]		
	Alipay (China) [491]	Geographical information	
	WeChat (China) [247]		

continued on next page

continued from previous page

	LeaveHomeSafe (China) [220]		
Decentralized	Stopp Corona (Austria) [359]	Encounter information	The malicious drivers can infer the trajectories of patients and match the trajectories with identifiers by creating multiple accounts and recording multiple routes. The geographical information is disclosed to the public for epidemic prevention.
	Stop COVID-19 (Croatia) [369]		
	eRouška (Czechia) [375]		
	Smittestop (Denmark) [209]		
	Koronavilkku (Finland) [327]		
	Corona-Warn-App (Germany) [331]		
	Apturi Covid (Latvia) [19]		
	Radar COVID (Spain) [334]		
	SwissCovid (Switzerland) [133]		
	NHS COVID-19 (United Kingdom) [442]		
	DP-3T (European) [404]		

continued on next page

continued from previous page			
	PACT (USA) [333]		
	Private Kit: Safe Path (USA) [143]	Geographical information	
	South Korea system [179]		
	HaMagen (Israel) [239]		

The centralized applications (e.g., TraceTogether (Singapore) [377], COVIDSafe (Australia) [87], and BlueTrace (Singapore) [48]) record the encrypted encounter history rather than the location information. Only the authority can decrypt the encounter history. The driver's identification can be obtained by the authority if the driver is infected. Other drivers can check their risk of infection by examining their encounter history. However, the patients' identities and trajectories are exposed to the authorities, which increases the risk of location privacy disclosure. The privacy-preserving capability of the applications can be decreased if other drivers collude to infer targets.

DP-3T (European) [404] and PACT (USA) [333] are two examples of decentralized applications. The two applications create a periodical key to generate several ephemeral identifiers. The ephemeral identifiers are broadcast as a beacon message within a region. The two applications store the received beacon messages with extra information. The applications match the ephemeral identifiers based on the information from the authority. Nevertheless, malicious drivers can infer the trajectories of patients and match the trajectories with identifiers by creating multiple accounts and recording multiple routes. Private Kit: Safe Path (USA) [143] and South Korea system [179] are the other types of COVID-19 applications. They periodically record the drivers' trajectories with time-stamped logs and pseudonyms based on the GPS (the South Korean system records detailed personal information). The drivers can report their trajectories to the authorities if they are infected. The authority can select information to be exposed to the public. The trajectories are disclosed to the public for epidemic prevention. Nevertheless, the applications cannot offer the expected location privacy-preserving capability, as everyone can view the trajectories.

The existing COVID-19 contract tracing applications balance location privacy protection capability and data utility by using anonymity [358]. The applications, e.g., TraceTogether (Singapore) [377] and COVIDSafe (Australia) [87], periodically broadcast random time-varying tokens as the driver's temporary IDs [79]. The applications record the encounter information of the drivers, which are employed to report infection risk [351]. However, the authorized entities can still link the temporary IDs and the driver's personal information (e.g., trajectory, phone number), even if the applications do not collect personal information on purpose [79].

2.3 Conclusion

In this chapter, we highlighted the threats of the existing localization techniques to location privacy protection. We then reviewed the existing LPPMs and discussed their advantages and limitations. The methods to balance the data utility and location privacy protection were researched in theoretical and practical scenarios.

The existing works in the field of location privacy preservation in vehicular networks have overlooked several crucial aspects. Firstly, many traditional methods primarily focus on generic 2D planar spaces and neglect the unique intricacies of road networks. These oversights result in impractical outcomes, often obfuscating locations that are off-road and thus irrelevant to vehicular navigation and data sharing. Furthermore, these approaches tend to fall short of providing effective privacy protection for drivers in real-world scenarios, where location privacy requirements can vary significantly depending on the context. Existing methods often lack the adaptability required to meet these varying privacy needs. Another critical gap in the existing research is the limited consideration of differential privacy, specifically in the context of road networks. The additive nature of differential privacy has been overlooked, making it challenging to detect illegal trajectories accurately. Detecting and mitigating malicious or unauthorized vehicle trajectories is essential in ensuring the reliability and security of vehicular networks. Furthermore, prior research has not sufficiently explored cooperative privacy protection mechanisms that leverage the collective power of multiple vehicles. Cooperative approaches have the potential to enhance privacy while maintaining data utility effectively. These collaborative methods can provide a higher level of protection against privacy breaches and collusion attacks. The integration of pseudonym swapping with obfuscation techniques remains underutilized in the existing literature despite its potential to improve privacy preservation. In summary, this thesis aims to address these research gaps by proposing a comprehensive differential privacy framework tailored to road networks, personalized location privacy-preserving mechanisms, and cooperative privacy protection strategies, ultimately advancing the state-of-the-art in location privacy for vehicular networks.

3

Personalized Location Privacy with RN-I

The proliferation of Location-based Services (LBS) leads to increasing concern about location privacy. Location obfuscation is a promising privacy-preserving technique but yet to be adequately tailored for vehicles in road networks. Existing obfuscation schemes are based primarily on the Euclidean distances and can lead to infeasible results, e.g., off-road locations.

In this chapter, we define Road Network-Indistinguishability (RN-I) to evaluate obfuscation-based location privacy-preserving schemes in road networks. To protect drivers' location privacy in road networks, we introduce a Personalized Location Privacy-Preserving (PLPP) scheme and prove it achieves RN-I. The PLPP scheme employs a dual-obfuscation algorithm, consisting of a connection perturbation and an interval perturbation, to obfuscate on-road locations. An efficient personalization algorithm is designed for the PLPP scheme to fine-tune location privacy budgets for capturing drivers' sensitive locations and privacy requirements. Experiments upon two real-world datasets confirm the location privacy-preserving capability, data utility, and efficiency of the introduced PLPP scheme.

3.1 Introduction

Obfuscation-based schemes can protect location privacy against untrusted LBS servers and eavesdroppers [219, 388, 31]. The idea is based on the local differential privacy technology where users locally perturb raw data with differential privacy schemes and employ the obfuscated data rather than the accurate data [402]. In obfuscation-based schemes, drivers generate obfuscated locations by adding controllable noise in their actual locations and report the obfuscated locations in LBS requests [321]. The obfuscated locations are selected according to the distances between the obfuscated and actual locations under the constraint of indistinguishability [31]. Although the obfuscation-based schemes compromise location accuracy, they can still be used in location-insensitive LBS, such as location-based recommendations [271].

Obfuscation-based schemes have not been rigorously developed for vehicles in road networks. In general-purpose obfuscation schemes, e.g., two-dimensional (2D) Laplace location privacy-preserving schemes, the distance between an actual location and the corresponding obfuscated location is measured using the Euclidean distance [31, 166, 388]. This, however, underestimates the distance between two locations on road networks because the route distance is no shorter than the Euclidean distance between any two locations [33]. Another issue of general-purpose obfuscation schemes is that they may generate off-road obfuscated

DOI: 10.1201/9781003518044-3

locations, for example, on a river [381]. Additional steps can be introduced to avoid off-road locations, which, however, increase the computational complexity [184].

The privacy-preserving levels of locations are expected to be fine-tuned across road networks to balance data utility and location-privacy protection [471]. Drivers can be sensitive to a small number of locations in road networks. If all locations are obfuscated at the highest level as the sensitive locations, the insensitive locations are over-obfuscated, and the data utility is penalized [414]. If only the sensitive locations are obfuscated, the sensitive locations are statistically different from the insensitive locations. This compromises location privacy [3].

In this chapter, we introduce a Personalized Location Privacy-Preserving (PLPP) scheme which provides customized location-privacy protection for vehicles in road networks. We start by defining Road Network-Indistinguishability (RN-I), where the distinguishability of locations is measured by the route distance. A connection-interval obfuscation algorithm is introduced to select on-road obfuscated locations according to privacy budgets and the route distances between obfuscated and actual locations. A personalization algorithm is designed to customize the privacy budgets of connections based on the sensitive locations specified by drivers.

This chapter presents the following key topics:

1) We define a new measure of location indistinguishability for road networks by applying the concept of differential privacy. The indistinguishability is measured only on road networks (as opposed to continuous 2D space). Route distances (i.e., distances along roads) are used to derive the differential privacy upper bound. Off-road locations are precluded from obfuscation operations.

2) We introduce a new dual-obfuscation algorithm that firstly probabilistically obfuscates an actual location to a connection and then obfuscates the location into a road interval between the actual location and the obfuscated connection. By carefully designing the probabilities, we prove that the dual-obfuscation design satisfies the new differential privacy-based definition of road network-indistinguishability.

3) Considering the non-uniform location privacy requirements of a driver at different locations, we apply the nearest neighbor interpolation to specify the privacy budgets for all locations based on the sensitive locations of the driver. The locations of the vehicle can be obfuscated consistently without exposing the sensitive locations, while improving the utility of the location data of less-sensitive locations.

We conduct comprehensive experiments on two real-world trajectory datasets and compare the introduced PLPP scheme with the existing 2D Laplace location privacy-preserving schemes [388, 166, 12]. Experimental results show that the PLPP scheme outperforms the 2D Laplace schemes in terms of data utility, privacy-preserving level, and efficiency.

The rest of this chapter is organized as follows. In Section 3.2, related works are reviewed, followed by the system model in Section 3.4. The introduced privacy-preserving scheme is presented in Section 3.5 and experimentally evaluated in Section 3.6, followed by conclusions in Section 3.7. Notation and abbreviations used in this chapter are listed in Table 3.1.

3.2 Related Work

Studies about cryptography, signature, trust mechanisms, LBS design, and game-theoretical models have been carried out to protect location data [419]. The scheme developed in [446]

TABLE 3.1: Summary of notation.

Notation	Description
$\mathcal{G}$	The graph transformed from a road network
$\mathcal{V}$	The set of connections on $\mathcal{G}$
$\mathcal{E}$	The edge set on $\mathcal{G}$
p/p'	The actual/obfuscated location
v/v'	The actual/obfuscated connection
$d_{\mathcal{G}}(p,p')$	The shortest route distance between p and p'
ϵ	Privacy budget
u_i	The i-th interval
α	The number of the intervals in a road
D	Connection degree
G_i	The i-th segment of $\mathcal{G}$
v_i^m	The i-th m-hop connection from sensitive locations
$W_{v_i^m}/W_{G_i}$	The weight of v_i^m/G_i
SL	The set of sensitive locations
sl_i	The i-th sensitive location
AEs	Adversary estimation Errors
r	The radius of the obfuscation region
δ	Weight threshold

protects both past and current LBS queries from each user with modified linkable spontaneous anonymous signature and the oblivious transfer technique. The scheme presented in [479] is an asynchronous localization protocol that constructs the relationship between propagation delay and location to eliminate the side-effect of asynchronous clock and mobility. A location-based recommendation is a popular LBS application and requires three types of information, i.e., user, activity, and location [95]. The location-aware recommendation scheme developed in [229] uses locality-sensitive hashing to hide the location information of both users and services. The scheme can provide accurate recommendation results while preserving users' location privacy.

Various schemes have been introduced to protect location privacy for vehicles, such as mix-zone [9], k-anonymity [476], and obfuscation [403]. In mix-zone schemes, drivers stop sending LBS queries in *mixing* areas and change their pseudonyms when leaving the areas [9]. Nevertheless, the adversary can link pseudonyms by analyzing the timing information and transition probabilities of the drivers. In k-anonymity schemes, drivers send a group of locations, including actual locations and phony locations, to LBS servers [476]. The k-anonymity schemes have a high communication cost because the drivers need to communicate with the k-nearest neighbors. Obfuscation-based schemes use perturbed locations rather than actual locations in location-insensitive LBS, such as location-based recommendations [127]. Drivers can select desired results from LBS returns based on their actual locations [35].

The obfuscation-based schemes are derived from Differential Privacy (DP) [426], which aims to protect an individual's private information while publishing aggregated data about a dataset or a single message [415, 134]. Popular obfuscation-based works are listed in Table 3.2, where we compare the scenarios and the methodologies of the schemes.

Based on differential privacy, Andrés *et al.* develop a scheme called Geo-Indistinguishability (Geo-I) [31]. The authors employ the Laplace scheme [200] to realize Geo-I on a two-dimensional (2D) plane. The Geo-I properly reports obfuscated locations

TABLE 3.2: Related obfuscation-based schemes.

Ref.	Scenario	Methodology	Personalized
PLPP	RN	RN-I	Yes
[12]	2D	Geo-I	Yes
[31, 166]	2D	Geo-I	-
[321]	RN	Geo-I	-
[388]	2D and RN	Geo-I	-

surrounding the actual location based on the obfuscation probability distribution when a driver requests LBS. The results show that the Geo-I can effectively protect location privacy under infrequent location updates [269]. The scheme developed in [166] reduces the computational consumption of the Geo-I on the 2D plane by introducing additional servers. A driver in the region covered by the additional servers only calculates obfuscated locations once and repeatedly reports the obfuscated location, so that the average computational consumption is almost zero. The scheme in [321] reduces the computational consumption of the Geo-I by dividing roads into intervals with a certain length. The locations in the same interval can be obfuscated in the same way, hence reducing the computational consumption.

Some obfuscation-based schemes focus on location privacy in road networks. The Graph-Exponential scheme (GEM), which satisfies the Geo-I, evaluates the privacy protection level and data utility of the traditional Geo-I in road networks [388]. The GEM sets the connections (such as turns, forks, and intersections) in a road network as the obfuscation candidates. The scheme obfuscates drivers' actual locations to connections directly based on the Geo-I. The scheme in [321] discretizes a road network with the same length intervals. The authors employ the route distance between two intervals to measure the indistinguishability of the Geo-I. Nevertheless, it is challenging to set the same length intervals in the road network. If the intervals are short to accommodate short roads, the computational consumption increases [415]. The correlation between the connections and privacy may be ignored if the intervals contain several short roads.

Various schemes protect drivers' sensitive locations at different levels to realize personalization. The personalized scheme in [438] measures the privacy requirements with personal attributes, e.g., access duration, frequency, and regularity. The scheme formulates an incomplete information game to balance quality-of-service and privacy protection. The movement regularity-based privacy requirement is studied in [487] for personalized pseudonym swapping. The scheme in [322] measures privacy requirements using intimacy, which specifies community edge density in social networks. Differential privacy and generative adversarial networks are employed to add noise to raw data. The scheme [85] specifies the personal privacy requirements of a location to be negatively correlated with the number of hops to sensitive locations. The algorithm in [157] orchestrates semantic privacy and location privacy based on drivers' requirements, which are measured according to the relationship between drivers. A game-theoretic model is then constructed in coupling with social-distance-based differential privacy to protect location privacy. The scheme in [12] designs the privacy requirement to be negatively correlated with the Euclidean distance between the current location and the last inferred location. The privacy requirement is used to calculate the privacy budget for obfuscation. This design reduces the exposure probability but requires the real-time calculation of the privacy requirement.

To the best of our knowledge, none of the existing personalized obfuscation schemes has considered the features of road networks, such as route distance and road network topology. Moreover, most of the existing personalized schemes require heavy computation based on historical behaviors. The introduced methodology aims to safeguard the location privacy of drivers within road networks. The PLPP scheme comprises two key components: a Connection-Interval Obfuscation algorithm and a Personalization algorithm. The Connection-Interval Obfuscation algorithm discretizes road edges into intervals and obfuscates intervals, rather than individual locations, to reduce computational overhead. This dual-obfuscation process, consisting of connection and interval perturbations, ensures location privacy while enhancing data utility. The Personalization algorithm empowers drivers to specify sensitive locations, allowing for varying levels of location privacy protection based on personalized privacy requirements. By calculating connection weights in a hop-by-hop manner, the algorithm customizes privacy budgets across the road network, ensuring sensitive locations receive the highest protection, while optimizing data utility for less sensitive areas. The PLPP scheme adheres to the introduced (ϵ,r)-Road Network-Indistinguishability, effectively safeguarding location privacy near sensitive locations and providing robust privacy protection in road network-based location-based services.

3.3 Adversary Model

Adversaries in this chapter are passive internal attackers or external attackers. The adversaries aim to infer the actual locations of drivers from the obfuscated locations in LBS requests. The adversaries are assumed to have full knowledge of drivers' obfuscation schemes and drivers' prior trajectories [321]. With an obfuscated location p_0' of a driver, the adversaries can derive the probability that the driver is actually at p_0 using the Bayes' Theorem, as given by

$$\Pr[p_0|p_0'] = \frac{\Pr[p_0'|p_0]\Pr[p_0]}{\int_{p\,\mathrm{in}\,\mathcal{G}} \Pr[p_0'|p]\Pr[p]dp}, \tag{3.1}$$

where p is a location in graph $\mathcal{G}$, and $\Pr[p]$ is the probability that the driver is at location p. $\Pr[p_0'|p]$ is the probability that the driver is actually at p and generates obfuscated location p_0' based on the obfuscation scheme.

For any location p_1 in $\mathcal{G}$, the adversaries can estimate how likely p_1 is the actual location, as given by

$$\Pr[p_1|p_0'] = \frac{\Pr[p_0'|p_1]\Pr[p_1]}{\int_{p\,\mathrm{in}\,\mathcal{G}} \Pr[p_0'|p]\Pr[p]dp}. \tag{3.2}$$

Then, the adversary selects the location $\hat{p}_0$, which is the most likely to be the actual location p_0, as the driver's actual location, i.e.,

$$\hat{p}_0 = \arg\max_{p_1\,\mathrm{in}\,\mathcal{G}} \Pr[p_1|p_0']. \tag{3.3}$$

Adversary estimation Error (AE) defines the shortest route distance between p_0 and $\hat{p}_0$, and evaluates the privacy protection effect of obfuscation-based schemes. As shown in Figure 3.1, the route distance between p_0 and p_0' is the shift distance, and the route distance between p_0 and $\hat{p}_0$ is the AE.

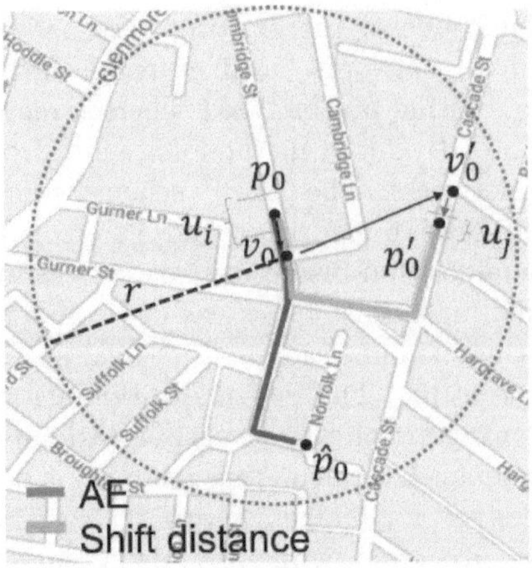

FIGURE 3.1: An example of the connection-interval obfuscation of the PLPP scheme.

3.4 System Model

In this section, we illustrate the road network and adversary model, as well as the important concept of this chapter.

3.4.1 Road Network Model

A road network can be viewed as a weighted directed graph $\mathcal{G} = (\mathcal{V}, \mathcal{E})$, where $\mathcal{V}$ is the set of connections (such as turns, forks, and intersections) and $\mathcal{E}$ is the edge set. Nodes in graph $\mathcal{G}$ are located at the same geographic positions as in the real world. The weight of an edge is the length of the corresponding road.

A vehicle can only move along an edge e, following the traffic rules at any given time. Drivers can access the LBS by sending their location information to the LBS servers. We consider the case that the LBS servers are not trusted and may disclose the location records of drivers to adversaries. A driver can conceal its actual location, i.e., p_0, by reporting an obfuscated location, i.e., p_0', using location obfuscation schemes.

3.4.2 Shift Distance

In this chapter, the definition of the data utility follows that in differential privacy, i.e., the difference between the obfuscated and actual location data, and measures the usability of the obfuscated location [10], [14], [36]. The data utility is higher, i.e., a better LBS response is provided, if the obfuscated location data used in the corresponding LBS request is closer to the actual location. In this sense, shift distance can be used to quantify the data utility of location obfuscation schemes. The smaller the shift distance is, the more accurate the LBS is and the higher the data utility is. Most existing location obfuscation schemes, such as [166] and [12], are designed for 2D planes (or maps) and employ the Euclidean distance between an actual location and its obfuscated location as the shift distance. In this chapter, we consider location obfuscation schemes in road networks and define the shift distance as

Definition 3.1 (Shift Distance) *Given an actual location p_0 and its corresponding obfuscated location p_0', the shift distance is the shortest route distance between the actual location p_0 and the obfuscated location p_0'.*

The data utility decreases with the increasing distance between an actual location and its obfuscated version, i.e., the shift distance. The shift distance has been extensively used in existing literature (e.g., [321] and [166]) as the only metric for evaluating the data utility.

3.4.3 Road Network-Indistinguishability

The Geo-I scheme, originally introduced for 2D Euclidean space, can be developed to protect location privacy in weighted and directed road networks. The Geo-I scheme perturbs actual locations using the concept of differential privacy, where a set of actual locations are geo-indistinguishable from the adversaries. We define the Road Network-Indistinguishability (RN-I) to measure on-road privacy-preserving schemes.

Definition 3.2 (Road Network-Indistinguishability) *A scheme satisfies (ϵ, r)-Road Network-Indistinguishability if and only if, for an obfuscated location p', any location pair p_i and p_j in the road network ($d_{min}(p_i, p_j) \leq r$) have*

$$\frac{\Pr[p_i|p']}{\Pr[p_j|p']} \leq e^{\epsilon d_{\mathcal{G}}(p_i,p_j)} \frac{\Pr[p_i]}{\Pr[p_j]}, \tag{3.4}$$

where $d_{\mathcal{G}}(p_i, p_j)$ and $d_{min}(p_i, p_j)$ are the shortest route distance and the Euclidean distance between p_i and p_j, respectively. r is the radius of the obfuscation region $\mathbb{R}^2$, the center of which is p_i. ϵ is the privacy budget as used in differential privacy [166]. A low value of ϵ corresponds to a high privacy-preserving capability.

By substituting (3.2) into (3.4), the latter can be transformed as

$$\Pr[p'|p_i] \leq e^{\epsilon d_{\mathcal{G}}(p_i,p_j)} \Pr[p'|p_j], \tag{3.5}$$

which indicates that p_i and p_j are RN-Indistinguishable, for any pair of p_i and p_j.

The route distance measures the shortest distances between two mutually accessible locations. The use of route distance does not introduce extra complexity to the introduced scheme. One reason is that the route distance can be pre-calculated and stored in vehicles. Another reason is that the route distances can help preclude off-road locations, hence reducing the computational complexity required for obfuscation.

3.5 Personalized Location Privacy-Preserving Scheme

Applying the RN-I, we introduce the Personalized Location Privacy-Preserving (PLPP) scheme to protect the location privacy of drivers in road networks. The PLPP scheme consists of a location obfuscation algorithm and a personalization algorithm. The obfuscation algorithm perturbs actual locations using the privacy budget calculated by the personalization algorithm.

3.5.1 Connection-Interval Obfuscation

There are two major types of road division schemes: (a) roads are divided into the same number of intervals, as shown in Figure 3.2(a); and (b) the intervals in the road network have the same length, as shown in Figure 3.2(b).

FIGURE 3.2: Roads are divided into intervals by (a) the same number of intervals; (b) the identical length of intervals.

The schemes (a) and (b) can be used to discretize and segment road networks. We choose method (a), because it can readily and meaningfully obfuscate a vehicle's location to somewhere on a different yet connected road. This obfuscated location preserves location privacy while remaining relevant to typical location-based services. The resolution (or granularity) of the segmentation is on a road basis. It is adequate and consistent with the requirements of location privacy and location accuracy for typical location-based services. While method (b) can also work under the introduced PLPP scheme, the method would excessively segment a typical urban road network. The granularity of the method could be excessively fine, as compared to our goal of obfuscating a vehicle's location to a different yet connected road. Furthermore, it could be tedious and computationally expensive to divide an urban road network into segments with equal length. This is due to the fact that the lengths of different roads can vary dramatically in a road network. The largest common factor is that road distances can be very short, resulting in an excessively large number of segments. It is worth pointing out that the method described in (b) can offer practical values on highways with long distances between entrances/exits. The introduced PLPP scheme can be applied to highways and obfuscate vehicles into different segments of a long highway by adopting the method.

The PLPP scheme discretizes every edge into intervals and obfuscates intervals rather than single locations to reduce computational consumption. We evenly divide each edge into α intervals (α is a certain number). The interval index can start from either side of the edge because interval u_i is obfuscated in the same way as $u_{\alpha-i}$.

Vehicles perturb actual locations by locally running the connection-interval obfuscation algorithm in the PLPP scheme, including connection and interval perturbations. The connection perturbation determines the privacy-preserving level of the PLPP scheme, while the interval perturbation obfuscates connections to intervals to improve data utility and enlarge AE. The pseudo-code of the connection-interval obfuscation algorithm is shown in Algorithm 1. An example is illustrated in Figure 3.1.

In Algorithm 1, steps 4 to 11 are the connection perturbation. An actual location p_0 in the interval u_i is first mapped to the nearest adjacent connection of u_i, denoted by v_0. Then, v_0 is perturbed to another connection v_0' within the obfuscation region $\mathbb{R}^2$ (centered at v_0 with radius r) based on the connection perturbation probability, i.e., $\Pr[v_0'|v_0]$, as given by

$$\Pr[v_0'|v_0] = \frac{1}{\sum\limits_{v \text{ in } \mathbb{R}^2} e^{-\frac{\epsilon}{2} d_\mathcal{G}(v_0,v)}} e^{-\frac{\epsilon}{2} d_\mathcal{G}(v_0,v_0')}, \tag{3.6}$$

where v is a connection in the obfuscation region $\mathbb{R}^2$.

The interval perturbation is given by steps 12 to 22 in Algorithm 1. Firstly, the PLPP scheme selects the connection v_1 that is adjacent to v_0' and along the shortest route between p_0 and v_0'. Then, the scheme selects an interval u_j on road (v_0', v_1) according to the interval

Algorithm 1 Connection-Interval Obfuscation Algorithm.

Input:
The road network, $\mathcal{G}$;
The segment privacy budget, ϵ;
An actual location, p_0;
The number of intervals, α;
The obfuscation radius of each connection, r.
Output:
The obfuscated location, p_0'.

$\triangleright$ **Setup**
$u_i \leftarrow (\mathcal{G},\ p_0)$ $\triangleright$ p_0 is in u_i
$v_0 \leftarrow (\mathcal{G},\ u_i)$ $\triangleright$ v_0 is u_i's nearest adjacent connection
$\mathbb{R}^2 \leftarrow (\mathcal{G},\ v_0,\ r)$ $\triangleright$ identify the obfuscation region $\mathbb{R}^2$
$\triangleright$ **Connection perturbation**
for v_i in $\mathbb{R}^2$ **do**
  Calculate $e^{-\frac{\epsilon}{2} d_{\mathcal{G}}(v_0,v_i)}$
end for
Calculate $\sum_{v \text{ in } \mathbb{R}^2} e^{-\frac{\epsilon}{2} d_{\mathcal{G}}(v_0,v)}$
for v_i in $\mathbb{R}^2$ **do**
  Calculate $\Pr[v_i|v_0]$ $\triangleright$ use (3.6)
end for
$v_0' \leftarrow (\mathbb{R}^2,\ \Pr[v_i|v_0])$ $\triangleright$ Select v_0' based on $\Pr[v_i|v_0]$
$\triangleright$ **Interval perturbation**
$v_1 \leftarrow (\mathcal{G},\ p_0,\ v_0')$ $\triangleright$ v_1 is adjacent to v_0' and on (p_0, v_0')
for u_k in road $(v_1, v_0',)$ **do**
  Calculate $e^{-\frac{|i-k|}{2\alpha}\epsilon}$
end for
Calculate $\sum_{k=1}^{\alpha} e^{-\frac{|i-k|}{2\alpha}\epsilon}$
for u_j in road $(v_1, v_0',)$ **do**
  Calculate $\Pr[u_j|u_i]$ $\triangleright$ use (3.7)
end for
$u_i' \leftarrow ((v_1,\ v_0'),\ \Pr[u_j|u_i])$ $\triangleright$ Select u_i' based on $\Pr[u_j|u_i]$
$p_0' \leftarrow (\mathcal{G},\ u_i')$ $\triangleright$ Randomly select a location in u_i'
return The obfuscated location p_0'.

perturbation probabilities, i.e., $\Pr[u_j|u_i]$, as given by

$$\Pr[u_j|u_i] = \frac{1}{\sum\limits_{k=1}^{\alpha} e^{-\frac{|i-k|}{2\alpha}\epsilon}} e^{-\frac{|i-j|}{2\alpha}\epsilon}, \tag{3.7}$$

which indicates that intervals u_i and u_j are (ϵ, α)-indistinguishable. Finally, the PLPP scheme randomly selects a location p_0' in the interval u_j as the obfuscated location. In the interval perturbation, only the indexes to the intervals are used so that intervals on different roads with the same index are perturbed following the same interval perturbation probability distribution.

The interval perturbation reduces the shift distance by $d_{\mathcal{G}}(v_0', p_0')$ and increases data utility compared to connection-only obfuscation schemes, e.g., the connection perturbation and the scheme in [388]. The shift distance, i.e., $d_{\mathcal{G}}(p_0, p_0')$, of the PLPP scheme is evaluated by

$$d_{\mathcal{G}}(p_0, p_0') = d_{\mathcal{G}}(p_0, v_0') - d_{\mathcal{G}}(v_0', p_0'). \tag{3.8}$$

The interval perturbation increases the AE and improves the location privacy-preserving effect on top of connection-only obfuscation schemes because adversaries can hardly infer the perturbed connections with on-road obfuscated locations.

Theorem 3.1 *The PLPP scheme satisfies the (ϵ,r)-RN-Indistinguishability that any two locations in an obfuscation region are (ϵ,r)-RN-Indistinguishable.*

Proof 3.1 *The connection perturbation in the PLPP scheme is formulated as* (3.6). *Let* $f(v_0) = \sum_{v\, in \mathbb{R}^2} e^{-\frac{\epsilon}{2} d_{\mathcal{G}}(v_0,v)}$. *With the actual location* p_0, *given another actual location* p_1 *and its nearest adjacent connection* v_1 *in* $\mathbb{R}^2$, *we have*

$$\frac{\Pr[v'|v_0]}{\Pr[v'|v_1]} = \frac{f(v_1)}{f(v_0)} e^{\frac{\epsilon}{2}(d_{\mathcal{G}}(v_1,v') - d_{\mathcal{G}}(v_0,v'))}, \tag{3.9}$$

where v_0 *is the nearest adjacent connection of the actual location* p_0 *and* v' *is the obfuscated connection.*

We next prove the triangle inequality of the road network.

If v_1 *is on the shortest route between* v_0 *and* v_2, *we have*

$$d_{\mathcal{G}}(v_0, v_2) - d_{\mathcal{G}}(v_1, v_2) = d_{\mathcal{G}}(v_0, v_1). \tag{3.10}$$

If v_1 *is not on the shortest route between* v_0 *and* v_2, *by assuming that* $d_{\mathcal{G}}(v_0, v_2) - d_{\mathcal{G}}(v_1, v_2) > d_{\mathcal{G}}(v_0, v_1)$, *we have*

$$d_{\mathcal{G}}(v_0, v_2) > d_{\mathcal{G}}(v_0, v_1) + d_{\mathcal{G}}(v_1, v_2),$$

which means that the route between v_0 *and* v_2 *is longer than the route between* v_0 *and* v_2 *through* v_1. *Then, the shortest route between* v_0 *and* v_2 *should pass* v_1. *Thus, the assumption is false. If* v_1 *is not on the shortest route between* v_0 *and* v_2, *the following inequality holds*

$$d_{\mathcal{G}}(v_0, v_2) - d_{\mathcal{G}}(v_1, v_2) \leq d_{\mathcal{G}}(v_0, v_1). \tag{3.11}$$

Combining (3.10) *and* (3.11), *the triangle inequality in road networks is proven.*

In other words, for any three connections (v_0, v_1, v_2) in a road network, $d_{\mathcal{G}}(v_0, v_2) - d_{\mathcal{G}}(v_1, v_2) \leq d_{\mathcal{G}}(v_0, v_1)$ holds.

Due to the triangle inequality in road networks $d_{\mathcal{G}}(v_0, v_1)$ *is no shorter than* $d_{\mathcal{G}}(v_1, v') - d_{\mathcal{G}}(v_0, v')$. *Then, we have*

$$\frac{\Pr[v'|v_0]}{\Pr[v'|v_1]} \leq \frac{f(v_1)}{f(v_0)} e^{\frac{\epsilon}{2} d_{\mathcal{G}}(v_0,v_1)}. \tag{3.12}$$

By employing the triangle inequality, we have $e^{-\frac{\epsilon}{2} d_{\mathcal{G}}(v_1,v)} \leq e^{-\frac{\epsilon}{2}(d_{\mathcal{G}}(v_0,v) - d_{\mathcal{G}}(v_0,v_1))}$. *Therefore,*

$$\sum_{v\, in \mathbb{R}^2} \left(e^{-\frac{\epsilon}{2} d_{\mathcal{G}}(v_1,v)} - e^{-\frac{\epsilon}{2}(d_{\mathcal{G}}(v_0,v) - d_{\mathcal{G}}(v_0,v_1))} \right) \leq 0, \tag{3.13}$$

which can be rewritten as

$$\sum_{v\, in \mathbb{R}^2} e^{-\frac{\epsilon}{2} d_{\mathcal{G}}(v_1,v)} - e^{\frac{\epsilon}{2} d_{\mathcal{G}}(v_0,v_1)} \sum_{v\, in \mathbb{R}^2} e^{-\frac{\epsilon}{2} d_{\mathcal{G}}(v_0,v)} \leq 0. \tag{3.14}$$

Based on the definition of $f(v_0)$, *we have*

$$f(v_1) - e^{\frac{\epsilon}{2} d_{\mathcal{G}}(v_0,v_1)} f(v_0) \leq 0. \tag{3.15}$$

Thus, the following inequality holds,

$$\frac{f(v_1)}{f(v_0)} \leq e^{\frac{\epsilon}{2}d_G(v_0,v_1)}. \tag{3.16}$$

Combining (3.12) *and* (3.16), *we have*

$$\frac{\Pr[v'|v_0]}{\Pr[v'|v_1]} \leq e^{\frac{\epsilon}{2}d_G(v_0,v_1)}e^{\frac{\epsilon}{2}d_G(v_0,v_1)} = e^{\epsilon d_G(v_0,v_1)}, \tag{3.17}$$

which satisfies (3.5). *As a result, the introduced connection perturbation achieves the RN-I.*

The interval perturbation is independent of the connection perturbation. This is because the interval perturbation obfuscates interval indexes rather than the outputs of the connection perturbation. Next, we prove the interval perturbation achieves (ϵ, α)*-indistinguishability.*

Let u_i *denote an actual interval (the i-th of* α *intervals), and* u_j *is another interval with the same or different sequence number of* u_i. u_k *is the perturbed interval.* u_i, u_j, *and* u_k *can be located on the same road or on different roads. Let* $g(u_i) = \sum_{k=1}^{\alpha} e^{-\frac{\epsilon}{2}\frac{|i-k|}{\alpha}}$. *Then,*

$$\frac{\Pr[u_k|u_i]}{\Pr[u_k|u_j]} = \frac{g(u_j)}{g(u_i)}e^{\frac{\epsilon}{2}(\frac{|j-k|}{\alpha}-\frac{|i-k|}{\alpha})}. \tag{3.18}$$

As $|a|-|b| \leq |a-b|$ *(a and b are two real numbers), we have* $|j-k|-|i-k| \leq |i-j|$. *Thus,*

$$\frac{\Pr[u_k|u_i]}{\Pr[u_k|u_j]} \leq \frac{g(u_j)}{g(u_i)}e^{\frac{\epsilon}{2}\frac{|i-j|}{\alpha}}. \tag{3.19}$$

As $|i-k|-|i-j| \leq |j-k|$ *holds, we have* $e^{-\frac{\epsilon}{2}\frac{|j-k|}{\alpha}} \leq e^{-\frac{\epsilon}{2}(\frac{|i-k|}{\alpha}-\frac{|i-j|}{\alpha})}$. *Therefore,*

$$\sum_{k=1}^{\alpha}\left(e^{-\frac{\epsilon}{2}\frac{|j-k|}{\alpha}} - e^{-\frac{\epsilon}{2}(\frac{|i-k|}{\alpha}-\frac{|i-j|}{\alpha})}\right) \leq 0, \tag{3.20}$$

which can be rewritten as

$$\sum_{k=1}^{\alpha} e^{-\frac{\epsilon}{2}\frac{|j-k|}{\alpha}} - e^{\frac{\epsilon}{2}\frac{|i-j|}{\alpha}}\sum_{k=1}^{\alpha} e^{-\frac{\epsilon}{2}\frac{|i-k|}{\alpha}} \leq 0. \tag{3.21}$$

Based on the definition of $g(u_i)$, *we have*

$$g(u_j) - e^{\frac{\epsilon}{2}\frac{|i-j|}{\alpha}}g(u_i) \leq 0. \tag{3.22}$$

Thus, the following inequality holds,

$$\frac{g(u_j)}{g(u_i)} \leq e^{\frac{\epsilon}{2}\frac{|i-j|}{\alpha}}. \tag{3.23}$$

Combining (3.19) *and* (3.23), *we have*

$$\frac{\Pr[u_k|u_i]}{\Pr[u_k|u_j]} \leq e^{\frac{\epsilon}{2}\frac{|i-j|}{\alpha}}e^{\frac{\epsilon}{2}\frac{|i-j|}{\alpha}} = e^{\epsilon\frac{|i-j|}{\alpha}}. \tag{3.24}$$

As a result, u_i *and* u_j *are* (ϵ, α)*-Indistinguishable. With the obfuscated interval* u_k, *the adversaries cannot distinguish the intervals or infer the obfuscated connection.*

Utilizing the PLPP scheme, the actual location is mapped to its nearest adjacent connection in the region $\mathbb{R}^2$. *Hence, when the connection perturbation satisfies the RN-I, locations in the region* $\mathbb{R}^2$ *are* (ϵ, r)*-Road Network-Indistinguishable.*

When an adversary obtains an obfuscated location in an LBS request, the adversary first estimates the perturbed connection. Then, the adversary infers the mapped connection with the obfuscated connection. Finally, the adversary tries to estimate the location with the inferred mapped connection and the obfuscated interval. The PLPP scheme can protect location privacy under this adversarial model since the scheme satisfies the RN-I, as given by Theorem 3.1.

3.5.2 Personalization Algorithm

Personalized location privacy protection can be achieved by protecting a driver's location data at different levels based on the driver's personal privacy requirements. For a driver, a relatively small number of locations are critical and sensitive, and need to be stringently protected. The other less sensitive locations also need to be protected in order to reduce the statistical difference between sensitive locations and less sensitive locations. On the other hand, if all locations are perturbed to the same extent as the sensitive locations, the utility of perturbed location data is compromised excessively. Thus, the privacy-preserving levels across locations should be carefully designed to protect sensitive locations while improving the location data utility of less sensitive locations.

The introduced PLPP scheme allows the drivers to specify sensitive locations and protects sensitive locations with the lowest privacy budget (i.e., at the highest level). Higher privacy budgets can be configured at less sensitive locations so that the location data at these locations can provide better utility. To achieve this, we introduce a personalization algorithm to tune the privacy budget ϵ of connections. The algorithm controls the perturbation level of the dual-obfuscation process. To be specific, a driver first inputs its sensitive locations to the introduced personalization algorithm for configuring the privacy budgets across the road network. The most sensitive locations have the highest weights and the lowest privacy budgets for the best possible privacy protection. The personalization algorithm calculates the weights of all connections hop-by-hop, starting from the sensitive locations. The algorithm exploits the idea of the nearest neighbor interpolation and captures unique road network features, e.g., route distance and road network topology. Finally, the introduced algorithm divides the road network into segments and calculates the privacy budget for each segment based on the weights of connections in the segment. By using this personalized privacy budget configuration, the introduced dual-obfuscation described in Section 3.5.1 can protect location privacy near sensitive locations as expected by the drivers and prevent exposing sensitive locations.

The road network is treated as a weighted directed graph $\mathcal{G}$. The road network is divided into g segments, i.e., $\mathcal{G} = \{G_1, G_2, \ldots, G_g\}$. The number and the size of segments can be adjusted based on drivers' privacy requirements. The introduced personalization algorithm first calculates the weights of all connections according to the most sensitive locations specified by a driver. The weight of a segment takes the maximum connection weight in the segment, and the privacy budget of the segment is the ratio of the initial privacy budget to the weight of the segment. The PLPP scheme uses the privacy budget of a segment as the private budget of the locations in the segment when obfuscating the locations.

The weights of the connections are calculated hop-by-hop, starting from the sensitive locations. Let $\mathcal{V}_m = \{v_1^m, v_2^m, \ldots\}$ denote the set of m-hop connections, where drivers can reach the i-th m-hop connection v_i^m via at least $(m-1)$ connections starting from a sensitive location. An example of $\mathcal{V}_1$ and $\mathcal{V}_2$ is given in Figures 3.3(a) and 3.3(b). The weight of v_i^m is denoted by $W_{v_i^m}$.

As shown in Algorithm 2, the calculation of connection weights consists of the following steps.

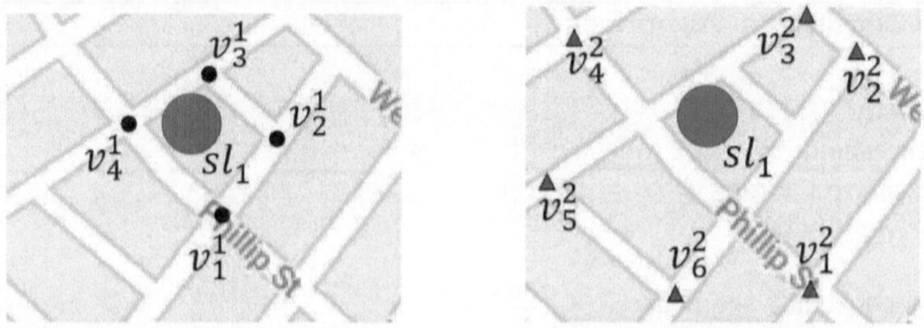

(a) The one-hop connection set $\mathcal{V}_1$.   (b) The two-hop connection set $\mathcal{V}_2$.

FIGURE 3.3: An example of $\mathcal{V}_1$ and $\mathcal{V}_2$, where sl_1 is a sensitive location.

- **Initialization (Steps 1-4):** The weights of sensitive locations and connections adjacent to the sensitive locations are set to 1. The weights of other connections are set to 0. The weight of 1 implies that the privacy budgets of sensitive locations are set to be initial privacy budget ϵ_0.

- **Weight Calculation (Steps 5-16):** The connection weights are calculated hop-by-hop, e.g., calculating the weights of the connections in $\mathcal{V}_3$ after finishing the calculation on $\mathcal{V}_2$. The weight calculation starts from $\mathcal{V}_2$ and ends when the weights of all connections have been calculated, or all the connection weights in $\mathcal{V}_m$ are less than δ.

  We employ the nearest neighbor interpolation[1] to calculate privacy budgets in the road network. Specifically, we design a new function, as given in (3.25), to quantify the weights a connection can copy from its one-hop neighbors. The weight of connection v_i^m, $m \geq 2$ depends on multiple $(m-1)$-hop connections that are adjacent to v_i^m. We first calculate all possible weights from the adjacent $(m-1)$-hop connections of v_i^m and then take the maximum one as $W_{v_i^m}$. The possible weight of v_i^m calculated from v_a^{m-1}, which is an $(m-1)$-hop connection and adjacent to v_i^m, i.e., $W_{v_i^m}^a$, is given by

$$W_{v_i^m}^a = \frac{\frac{D_{\max} - D_{v_i^m} + 2}{d_{\mathcal{G}}(v_a^{m-1}, v_i^m)} \times W_{v_a^{m-1}}}{\sum\limits_{v_j^m \in \mathcal{A}_a^m, j \neq i} \frac{1}{d_{\mathcal{G}}(v_a^{m-1}, v_j^m)} + \frac{D_{\max} - D_{v_i^m} + 2}{d_{\mathcal{G}}(v_a^{m-1}, v_i^m)}}, \tag{3.25}$$

  where $D_{v_i^m}$ is the degree of connection v_i^m (the sum of in-degree and out-degree); $D_{\max}$ is the maximum degree of connections in the segment; $\mathcal{A}_a^m$ is a set of m-hop connections that are adjacent to connection v_a^{m-1}. As road networks can be treated as strong-connected directed graphs, the degrees of the connections are at least two. $W_{v_i^m}$ can take the maximum one across different $W_{v_i^m}^a$, i.e., $W_{v_i^m} = \max(W_{v_i^m}^a)$.

- **Finalization (Steps 17-21):** The weights of connections which are less than δ are set to δ. The weight of a segment, i.e., W_{G_i}, takes the maximum weight of connections in the segment, and the privacy budget of the segment is calculated based on its weight, i.e., $\epsilon_{G_i} = \frac{\epsilon_0}{W_{G_i}}$.

There are two key parameters in (3.25), i.e., the degree of the connection $D_{v_i^m}$ and the shortest route distance to the last hop connections, $d_{\mathcal{G}}(v_a^{m-1}, v_i^m)$. Firstly, low-degree

[1]The general nearest neighbor interpolation in image processing infers the pixel of a point by copying the pixels of its nearest neighbors.

Algorithm 2 Personalization Algorithm.

Input:
The road network, $\mathcal{G}$;
The weight threshold, δ;
The set of sensitive locations, $SL = \{sl_1, sl_2, \cdots, sl_n\}$;
The initial privacy budget, ϵ_0.
Output:
Privacy budget for each segment.

▷ **Initialization**
$\mathcal{V}_m, m = 1, 2, \cdots, M.$ ▷ Initialize connection sets
$W_{SL} \leftarrow 1$ ▷ Set the weights of sensitive locations
$W_v \leftarrow 1$, for all $v \in \mathcal{V}_1$
$W_v \leftarrow 0$, for all $v \in \mathcal{V}_i$, $i > 1$
▷ **Weight calculation**
$m = 2$
while $m \leq M$ **do**
  for $v \in \mathcal{V}_m$ **do**
    Calculate W_v ▷ With (3.25)
  end for
  if $\forall v \in \mathcal{V}_m$ has $W_v < \delta$ **then**
    break
  else
    $m++$
  end if
end while
$W_v \leftarrow \delta$, if $W_v < \delta$
▷ **Finalization**
for Segment G_i in road networks **do**
  $W_{G_i} \leftarrow \max_{v \text{ in } G_i} \{W_v\}.$
  Calculate $\epsilon_{G_i} = \frac{\epsilon_0}{W_{G_i}}$ ▷ The segment privacy budget
end for
return The privacy budget for each segment ϵ_{G_i}

connections imply more explicit travel directions than high-degree connections because low-degree connections offer fewer choices. Thus, low-degree connections should be protected better than high-degree connections. Secondly, connections near sensitive locations are expected to be protected better than connections that are far away from sensitive locations.

The introduced personalization algorithm customizes privacy budgets for connection-interval obfuscation according to drivers' sensitive locations and the features of road networks. In contrast, existing works (e.g., [388]) use a uniform private budget across a road network. The personalization algorithm in the PLPP scheme runs once when drivers configure sensitive locations and thus does not slow down the connection-interval obfuscation.

Given a road network and sensitive locations, the connection-interval obfuscation can use the private budget ϵ, the obfuscation radius r, and the number of intervals per road α, to perturb locations. The private budget ϵ is calculated using the introduced personalization algorithm with the initial obfuscation privacy budget ϵ_0 and the connection weight threshold δ. A driver can use a small ϵ_0 and/or a large r for large AEs and long-shift distances. The driver can also configure a large α for fine interval perturbation.

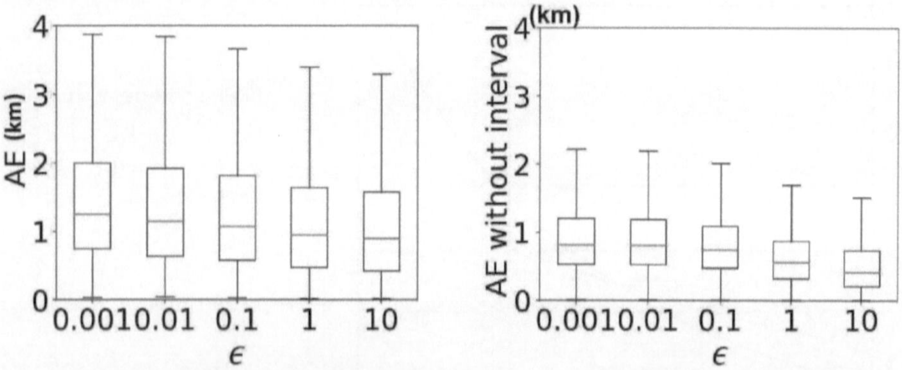

(a) AE of the PLPP scheme

(b) AE of the scheme in [388]

FIGURE 3.4: The comparison of the AE of the introduced PLPP scheme and the existing scheme.

3.6 Experimental Results

In this chapter, we evaluate the PLPP scheme with two real-world trajectory datasets, i.e., GeoLife GPS Trajectories (182 drivers with 17,621 trajectories in Beijing, China) [486] and T-Drive trajectory (10,357 divers in Beijing, China) [462].

3.6.1 Location Privacy Protection

We first compare the AEs of the PLPP scheme and a connection-only scheme, i.e., [388]. We select the vehicle trajectories from the two datasets within the same region.[2] We build a simulation road network with clear connections in the region and then vertically project trajectories onto the nearest roads if they are not on the simulation road network. The road network is evenly divided into ten segments. The privacy budgets of the segments are calculated using Algorithm 2 with ten random sensitive locations on trajectories and the initial privacy budget ϵ.

The experimental results reveal that the PLPP scheme protects location privacy better than the connection-only scheme [388], as shown in Figure 3.4. The boxplots indicate that the AEs of the PLPP scheme are evenly distributed across longer ranges than the AEs of the connection-only scheme. The interquartile ranges in Figure 3.4(a) are almost twice as large as those in Figure 3.4(b) because adversaries cannot infer the obfuscated connections after the interval perturbation in the PLPP scheme. The medians of the AEs of the PLPP scheme are longer than those of the scheme in [388] across all ϵ. The median is about double under the PLPP scheme (1 km) than it is under the connection-only scheme (0.5 km) when $\epsilon = 10$. The medians in the two subfigures decrease with the increase of ϵ, indicating that the drivers can set small ϵ for high location privacy protection.

The indistinguishability of the PLPP scheme, i.e., $\frac{\Pr[v'|v_0]}{\Pr[v'|v_1]} \leq e^{\epsilon d_{\mathcal{G}}(v_0,v_1)}$ in (3.17), is validated in Figure 3.5. In this experiment, the obfuscation radius is 400 m. We select two actual locations v_0 and v_1 when $d_{\mathcal{G}}(v_0, v_1) = 1$ km and run the PLPP scheme for the obfuscated locations v'. As shown in Figure 3.5, the left-hand side of (3.17) is bounded, validating the

[2]The longitude and latitude of the region are in ranges of [116.3564, 116.3740], and [39.907, 39.9241], respectively.

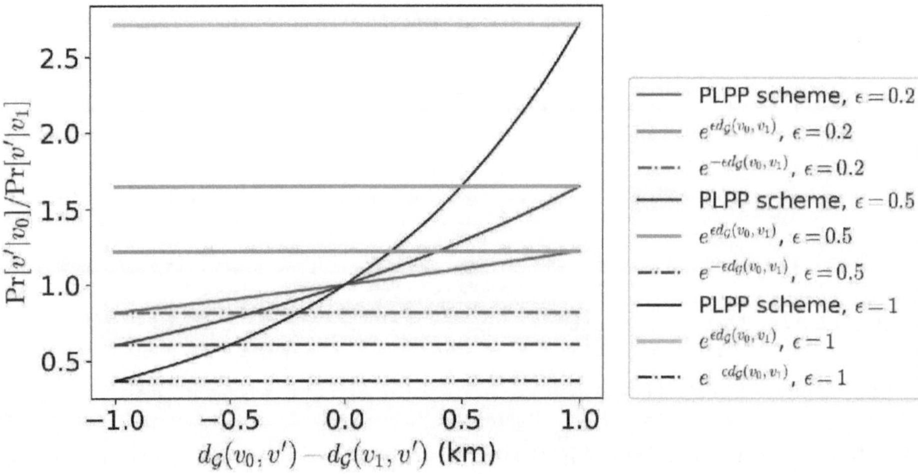

FIGURE 3.5: The indistinguishability of the PLPP scheme. The obfuscation radius is 400 m.

indistinguishability of the PLPP scheme. An adversary can hardly distinguish actual locations that have a similar route distance from the obfuscated location, i.e., $\frac{\Pr[v'|v_0]}{\Pr[v'|v_1]} = 1$, with $d_{\mathcal{G}}(v_0, v') - d_{\mathcal{G}}(v_1, v') = 0$. We also see that the left-hand side of (3.17) is approximately linear under a small ϵ, e.g., $\epsilon = 0.2$. The reason is that when x is much smaller than one, $e^x = 1 + x + \frac{x^2}{2!} + \frac{x^3}{3!} + \cdots + \frac{x^n}{n!} \approx 1 + x$.

The influence of the obfuscation radius on the shift distance and AE is shown in Figure 3.6. The obfuscation radius ranges from 300 m to 700 m. The other parameters are the same as those in Figure 3.5. All the connections are in the same segment. The initial privacy budgets are 0.001 and 1 to simulate a strict privacy-preserving case and a mild privacy-preserving case, respectively. The shift distance and AE grow approximately linearly with the obfuscation radius. As a result, drivers need to trade data utility for location privacy, but the AEs are much larger (almost double) than the shift distances. The growth is approximately linear because the route distance between actual locations and obfuscated location candidates linearly increases with the rising obfuscation radius in even road net-

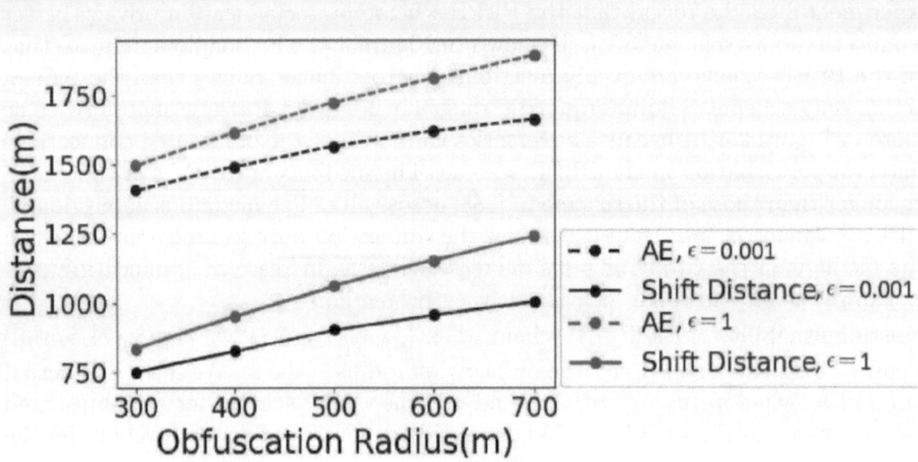

FIGURE 3.6: The influence of the obfuscation radius.

works. The shift distance and AE increase more slowly under $\epsilon = 1$ than under $\epsilon = 0.001$. This is because the obfuscated locations under big ϵ are closer to the actual locations than the obfuscated locations under smaller ϵ.

We proceed to compare the PLPP scheme with a popular 2D location obfuscation scheme, i.e., the planar Laplace scheme [166, 12]. The planar Laplace scheme treats a road network as a 2D plane and employs the Euclidean distance metrics. Figure 3.7 compares the average AE and shift distance of the PLPP scheme and the planar Laplace scheme with GeoLife GPS Trajectories. For comparison purposes, all the segments use the same privacy budget ϵ in the PLPP scheme.

As shown in Figure 3.7, the PLPP scheme can achieve a longer average AE and shorter average shift distance than the scheme in [166], especially in the case of high-privacy budgets. The gap between the two schemes grows with the increasing privacy budget. The PLPP scheme can achieve a shorter average shift distance when ϵ is higher. The average shift distance measured by the route distance can be longer than the obfuscation radius because the route distance is no shorter than the Euclidean distance. The PLPP scheme has a similar average shift distance to the 2D Laplace scheme under $\epsilon = 0.001$. The reason is that locations are obfuscated with similar probabilities in the two schemes when the privacy budget is 0.001. When the privacy budget is 10, the average shift distance and the average AE with a 300 m radius are similar to those with a 600 m radius. This is because the obfuscated locations are close to actual locations in the case of low-privacy requirements (high ϵ). The difference between shift distances under the two radii also decreases with the

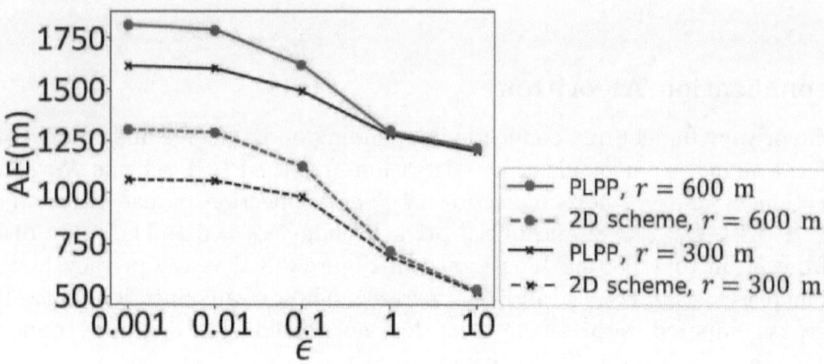

(a) Comparison of AE

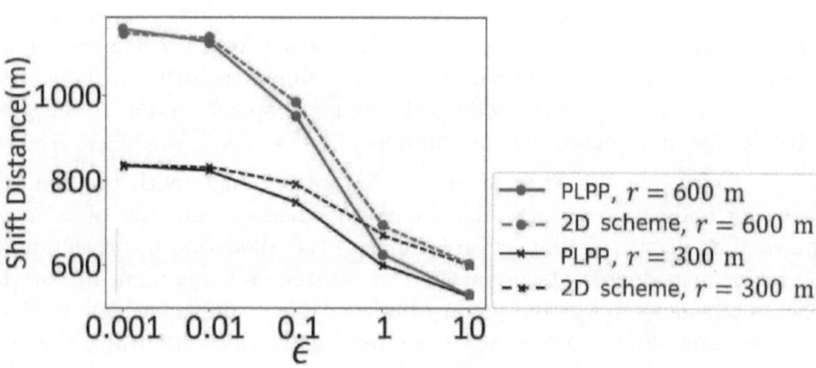

(b) Comparison of shift distance

FIGURE 3.7: A comparison of the PLPP scheme and 2D Laplace scheme.

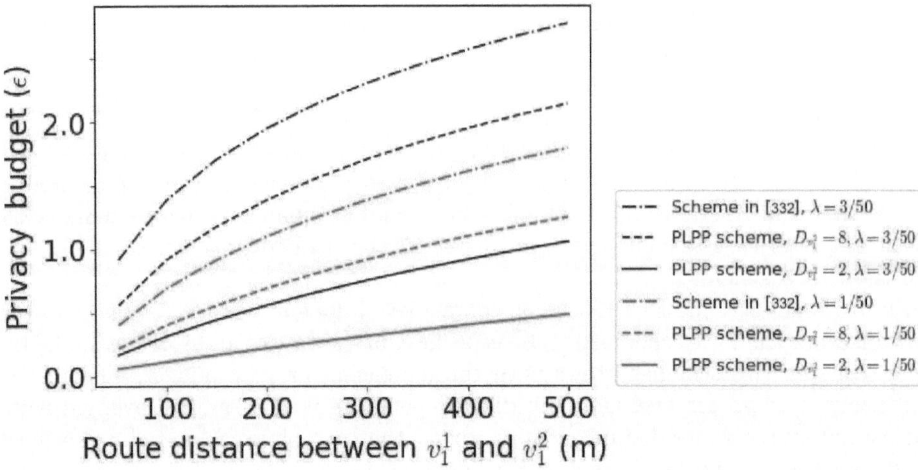

FIGURE 3.8: The impact of the degree of a single connection and route distance to a sensitive location in the personalization algorithm.

increasing privacy budget, as shown in Figure 3.7(b). This is because both schemes are likely to select close locations as the obfuscation results under high-privacy budgets. The shift distance of the PLPP scheme is shorter than the 2D planar Laplace scheme in road networks.

3.6.2 Personalization Algorithm

We evaluate the privacy budget of a connection depending on its degree and route distance to a sensitive location in the personalization (Algorithm 2) of the PLPP scheme. We consider a 1-hop connection v_1^1 and a 2-hop connection v_1^2. The connection v_1^1 has other adjacent connections v_i^2 ($i \geq 2$). $D_{\max} = 8$, and initial privacy budget $\epsilon_0 = 0.1$. The route distance $d_{\mathcal{G}}(v_1^1, v_1^2)$ is from 50 m to 1,000 m. It is shown in Figure 3.8 that the privacy budget of connection v_1^2 increases, as $d_{\mathcal{G}}(v_1^1, v_1^2)$ and $D_{v_1^2}$ increase. The ϵ of a connection in the PLPP scheme is lower, as compared to the scheme that does not consider the degree of connections (i.e., setting $(2 + D_{\max} - D_{v_i^m})$ in (3.25) to 1).

We then evaluate the privacy budget of multiple connections on a continuous road affected by the number of connections, the route distance to the first connection v_1^1, and the degree of connections. We consider a 1000 m road starting from a 1-hop connection on which five or eleven connections are evenly spaced and evaluate the privacy budget of each connection. The m-th connection on the road is the first m-hop connection v_1^m ($1 \leq m \leq 5$ or $1 \leq m \leq 10$). Other parameters are as follows, $D_{\max} = 10$, $\sum_{i \geq 2} \frac{1}{d_{\mathcal{G}}(v_1^j, v_i^{j+1})} = \frac{1}{100}$ ($1 \leq j \leq m-1$), and $\epsilon_0 = 0.1$. It is shown in Figure 3.9 that, with the same route distance, the privacy budget of a connection increases when the number of connections on the road grows. The degree of connection influences the allocation of privacy budgets. If a connection has a high degree, its privacy budget increases faster than its low-degree counterparts. Some schemes set a global privacy budget [445] or only consider the distance between two connections [401]. Compared to our personalization algorithm, the privacy budgets of the schemes developed in [401] and [445] increase faster, as shown in Figure 3.9. Compared with the schemes (e.g., [445, 401]), which do not capture the connection degree, the introduced personalization algorithm ensures that a connection can inherit more privacy protection from the sensitive location.

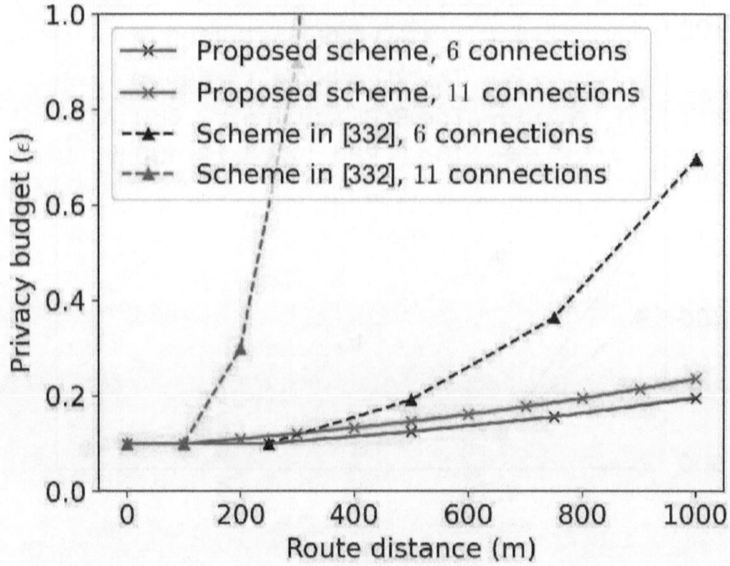

FIGURE 3.9: The privacy budget of connections on a road impacted by the number of connections, the route distance to the first connection v_1^1, and the degree of each connection.

A comparison study is carried out between the introduced personalization algorithm and the personalization algorithm developed in [12]. The scheme in [12] sets the privacy budgets according to the distance between the current location and the last inferred location. We randomly select 20 trajectories in the simulated road network, where each trajectory consists of 10 to 20 actual locations. The privacy budgets calculated with the two personalization algorithms are combined with the introduced connection-interval perturbation scheme. The obfuscation radiuses of 400 m and 600 m are used. As shown in Figure 3.10, the introduced personalization algorithm achieves shorter shift distances (i.e., better data utility) than the existing algorithm developed in [12], especially for small ϵ_0. For example, the obfuscation with the introduced personalization algorithm provides a shift distance of 700 m when $\epsilon_0 = 0.1$ and $r = 400$ m. In contrast, the obfuscation with the algorithm developed in [12] achieves a shift distance of over 1,100 m. The shift distance of the PLPP scheme decreases faster than that of the existing algorithm, indicating that the introduced personalization algorithm can improve data utility. The shift distances of the two algorithms are similar, when the privacy budget is high, e.g., $\epsilon_0 = 10$. This is because the two schemes tend to select connections close to the actual locations to be obfuscated connections when data utility is preferred over privacy.

3.6.3 Influence of the Road Network

We generate virtual road networks with different node densities to show the impact of road networks on the PLPP scheme. The node (connection) densities are set to $80/\text{km}^2$, $40/\text{km}^2$, and $20/\text{km}^2$. The trajectories are from the T-Drive dataset. We select 100 trajectories randomly and map the trajectories in the generated road networks.

An adversary has smaller AE in the high node density regions because locations estimated by the adversary are closer to the actual locations. With $80/\text{km}^2$ node density, the average AE of the PLPP scheme is the shortest among the three given node densities. In Figure 3.11(a), we see that the AE in the network with the node density of $40/\text{km}^2$, when $\epsilon = 0.001$ is slightly higher than the AE in the network with the node density of $20/\text{km}^2$.

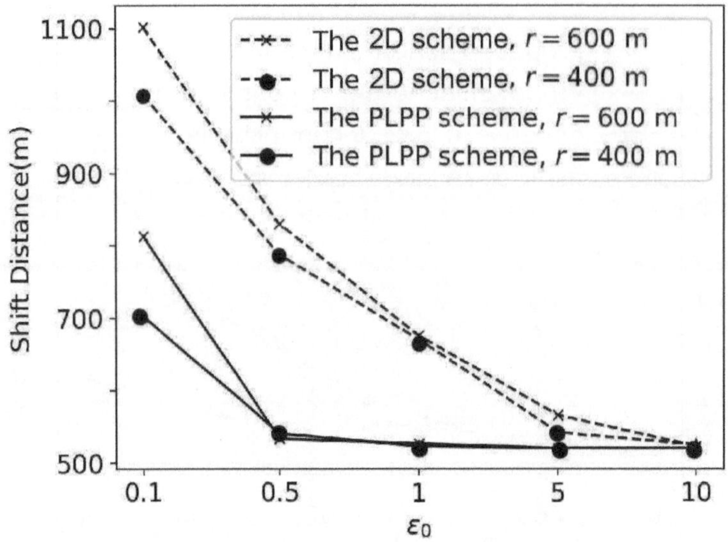

FIGURE 3.10: A comparison of the personalization algorithms in the PLPP scheme and 2D scheme.

The reason is that the obfuscated region covers a limited number of obfuscation candidates when the node density is low. The limited connections may not provide the expected privacy-preserving capability.

The PLPP scheme can provide better data utility in high-density road networks than it does in low-density road networks, as shown in Figure 3.11(b). The route distances between obfuscation candidates and actual locations in high-density networks are shorter than in low-density networks. Thus, the expected average shift distances in the high-density regions are shorter than those in the low-density regions. According to Figures 3.11(a) and 3.11(b), the curves with different node densities show similar trends, as ϵ increases.

The impact of the interval number α on the interval selection probability is evaluated in Figure 3.11. The number of intervals per road, i.e., α, can be configured using machine learning techniques, e.g., [128] and [472], and evolutionary computing algorithms, e.g., [413]. In this chapter, we configure α based on Taguchi's method [128], which reduces experimental tests by separating orthogonal parameters. In the PLPP scheme, only α and ϵ are used

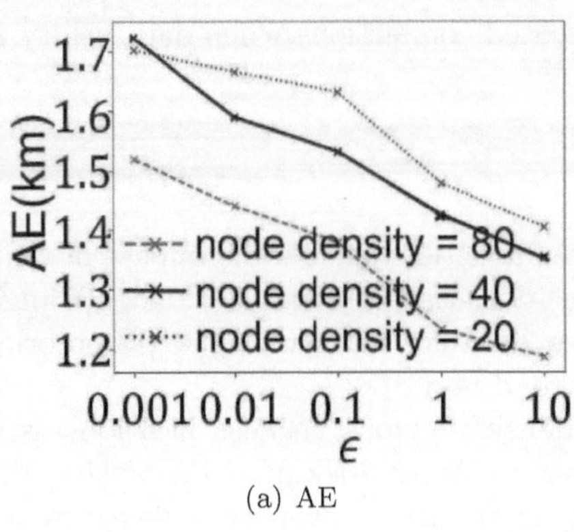

(a) AE

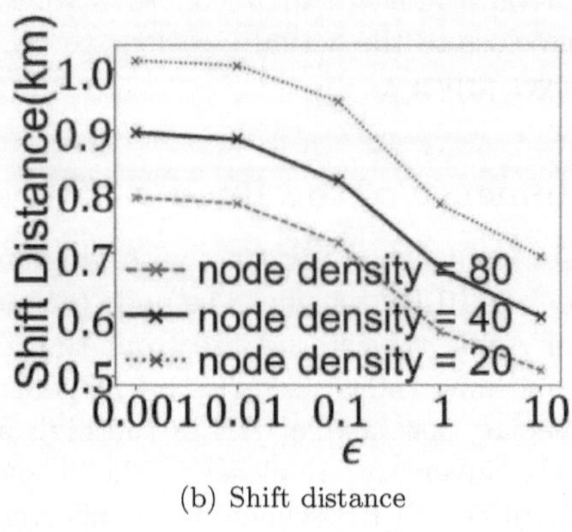

(b) Shift distance

FIGURE 3.11: The impact of the node density on AE and shift distance.

in the interval perturbation and orthogonal to other parameters, e.g., r. In Figure 3.11, $\alpha \in \{15, 20\}$ and $\epsilon = 1$. The length of the road is 90 m, where the driver is 13 m away from the connection, and its actual interval index is 3 for both $\alpha = 15$ and 20. The obfuscation probability is horizontally symmetric since the interval u_i is obfuscated in the same way as $u_{\alpha-i}$; see (7).

3.6.4 Running Time

The PLPP is designed to minimize the latency due to obfuscation by enabling privacy budgets, the connection perturbation distribution, and the interval perturbation distribution to be pre-calculated. Before any LBS requests, the PLPP scheme can calculate privacy budgets for all connections and then build constant connection perturbation and interval perturbation tables. Each entry of the tables gives the perturbation probability of a pair of connections (3.6) or intervals (3.7). The sizes of the connection perturbation and interval perturbation tables are $O(n^2)$ and $O(\alpha^2)$, respectively, where n is the number of connections in the road network. During the LBS requests, the connection-interval obfuscation can look up the tables to probabilistically select the obfuscated connections and locations.

We compare the running time of the introduced PLPP scheme with the personalization algorithm in [12] and the non-personalized obfuscation algorithm in [388], as shown in Table 3.3. The personalization algorithm developed in [12] sets privacy budgets based on the distance between the current and the last inferred locations. The obfuscation algorithm presented in [388] employs a 2D single-obfuscation and maps the obfuscated locations to the nearest connections in the road network. To show the running time of each module, we combine our dual-obfuscation with the personalization algorithm in [12] and our personalization algorithm with the 2D single-obfuscation [388]. The other parameters are as follows: 500 actual locations are randomly selected, the obfuscation radius r is 500 m, and the initial privacy budget ϵ_0 is 1.

As shown in Table 3.3, the introduced PLPP scheme is faster than the dual-obfuscation with the personalization algorithm in [12] in all cases with up to 19% improvement (i.e.,

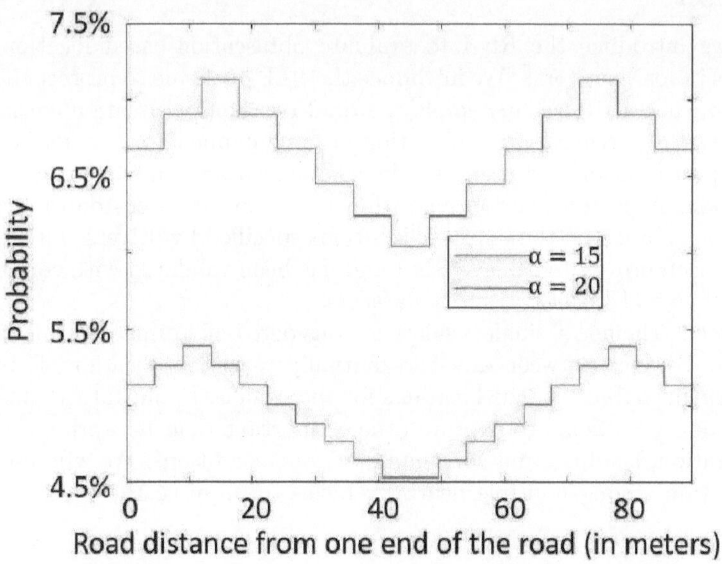

FIGURE 3.12: The impact of the interval number α on interval selection, where α is set to 15 and 20. The privacy budget ϵ is set to 1.

TABLE 3.3: Comparison of running time (ms).

Road Density (connections/km^2)	The PLPP Mechanism	The Interval Perturbation in the PLPP Mechanism	The Personalization Algorithm in [12]	The Mechanism in [388]
100	0.1954	0.0060	0.2275	0.2772
150	0.3091	0.0059	0.3670	0.5426
200	0.4029	0.0059	0.4947	0.7201
250	0.4907	0.0060	0.6083	0.8059

0.4907 ms vs. 0.6083 ms). The introduced personalization algorithm only runs once to calculate the privacy budgets for all connections when drivers configure their sensitive locations before obfuscation. In contrast, the personalization algorithm in [12] needs to be executed for each obfuscation. The PLPP scheme is also faster than the 2D single-obfuscation [388]. This is because the 2D single-obfuscation [388] needs to map the obfuscated locations onto the road network, which is time-consuming. Table 3.3 also shows that the running time of the interval perturbation is only about 0.006 ms. Compared to that of the whole PLPP scheme, the overhead of the interval perturbation is negligible. Moreover, the running time of the interval perturbation does not grow with the connection density. The reason is that the small constant α (i.e., 5 in the experiment) and the same interval perturbation distribution are applied to all roads.

3.7 Conclusion

In this chapter, we introduce the RN-I to evaluate obfuscation-based location privacy-preserving schemes in road networks. We introduce the PLPP scheme to protect the location privacy of vehicles. The PLPP scheme employs a dual location obfuscation consisting of a connection and interval perturbation. A location is first mapped to a remote connection in the connection perturbation and then pseudo-randomized around the connection in the interval perturbation. A personalization algorithm is developed to customize the privacy budgets of connections based on the sensitive locations specified by drivers. The introduced PLPP scheme has been proven to achieve RN-I and has been validated with comprehensive experiments using two real-world trajectory datasets.

In the introduced scheme, a single vehicle is considered in terms of its location and privacy. The route distance between any two mutually reachable locations is utilized to evaluate the indistinguishability of the locations for the vehicle. In our future work, we will generalize the introduced scheme to capture other important road network features, such as velocity, acceleration, heading, and surrounding traffic conditions. We will also consider potential collaboration among multiple nearby vehicles to preserve their location privacy cooperatively.

4

Vehicle Trajectory Obfuscation and Detection

In this chapter, we first introduce a Convolutional Neural Network (CNN) based detection mechanism to detect illegal trajectories without requiring the drivers' actual location. The introduced scheme has high accuracy in detecting illegal trajectories even if the drivers protect actual location data with various noise sizes.

The rest of this chapter is organized as follows. Sections 4.1 and 4.2 study the background of vehicular trajectory protection and detection and the existing works of vehicle trajectory obfuscation and detection, respectively. Section 4.3 describes the introduced trajectory detection mechanisms. Section 4.4 evaluates the introduced trajectory detection mechanisms with real-world road network datasets. Section 4.5 concludes the introduced vehicular trajectory protection and detection mechanism.

4.1 Introduction

Location-based Services (LBS) have been extensively and deeply developed as an essential part of smart cities to provide various services, e.g., traffic analysis and urban planning [394, 355]. Smart cities inevitably require the spatio-temporal location data of vehicles, which is highly correlated with a driver's private information (e.g., home address, company address, and religion) [422]. LBS providers rely on location data from drivers to offer various services [241]. However, adversaries can infer drivers' personal information by analyzing location data [439]. It is necessary to protect location data in Internet-of-Vehicles (IoV) for drivers' privacy while ensuring high Quality of Service (QoS).

Obfuscation schemes [32] and adaptive location privacy-preserving schemes [12] have been developed to protect location privacy. Obfuscation schemes add noise to obfuscate drivers' actual location data, which reduces the data utility of the location data. The obfuscated location data are indistinguishable from each other, which leads to the fact that LBS providers would collect illegal location data. The location privacy-preserving schemes aim to balance the QoS of LBS and privacy-preserving capability by analyzing the drivers' requirements.

Malicious drivers breach legal drivers' profit. For example, malicious drivers can occupy more benefits than they deserve by deliberately modifying their trajectory data in Taxi service [4, 132]. Malicious drivers also use location privacy-preserving schemes to protect their location data. By analyzing location data, which concludes illegal location data, smart city applications cannot provide an acceptable QoS of LBS. Therefore, LBS should detect illegal location data to ensure high QoS. If the malicious drivers employ location privacy-preserving schemes (e.g., obfuscation schemes) as the legal drivers, detecting illegal data

DOI: 10.1201/9781003518044-4

becomes difficult.

We study illegal location data detection and introduce a personalized obfuscation scheme and an illegal trajectory detection mechanism. This chapter has two-fold topics as follows:

- We employ differential privacy in the introduced personalized obfuscation scheme to adaptively protect drivers' location privacy and provide high QoS of LBS in road networks.
- We introduce a CNN-based detection mechanism to detect illegal trajectories without requiring the drivers' actual location. The introduced scheme has high accuracy in detecting illegal trajectories even if the drivers protect actual location data with various noise sizes.

We conduct experiments with the real-world road network dataset extracted from Open Street Map (OSM)[1] to evaluate the introduced scheme.

The illegal trajectories are generated with fake speeds and paths with the real-world road network dataset. We also evaluate the introduced scheme in the case that the drivers adaptively protect their location data, i.e., using different privacy levels. The experimental results show that the introduced detection scheme achieves at least 94% Area Under the Curve (AUC) score when detecting the illegal obfuscated trajectory.

4.2 Related Work

The existing smart city provides location-based services by mining trajectory data that are transmitted in vehicular networks [154]. Wang et al. have studied the privacy challenges in smart cities and analyzed the privacy leakage in LBS [419]. The authors pointed out that the smart city can provide high-quality services if trajectory privacy is well protected.

The previous studies of location obfuscation mechanisms perturb a driver's actual location and report an obfuscated version to LBS. Derived from the differential privacy [102], scheme in [32] first developed the concept of geo-indistinguishability. The scheme follows the idea of geo-indistinguishability and uses Laplace distribution to add controlled noise to protect location data locally. Yu et al. [458] improved the two-phase dynamic differential location privacy scheme by integrating the inference error expectation and geo-indistinguishability [363]. The improved framework effectively protects location privacy in a 2D map. The authors developed an adaptive location privacy-preserving mechanism in [12] to balance location privacy and utility. The mechanism calculates the amount of noise before adding noise to actual location data. The calculation is based on the correlation level between the driver's current location and the previous obfuscated locations. With the concept of differential privacy, Xiao et al. [434] improved a location-cloaking system to protect drivers' location data in a 2D map. The obfuscation locations generated by the existing 2D obfuscation mechanisms might be located at unreachable locations, e.g., in the river, which breaches the location privacy-preserving capability of the obfuscation mechanisms.

The existing illegal trajectory detection mechanisms are classified into machine-learning-based detection and rule-based detection. The illegal trajectory data is detected by utilizing GPS data in the rule-based detection mechanisms. Machine-learning-based detection mechanisms classify trajectory data as legal and illegal by using techniques like deep neural

[1]Open Street Map is an open-source database of the world's geographic map. https://www.openstreetmap.org/

networks. Chen et al. [76] improved an efficient real-time trajectory detection method with low processing overhead. The method uses the window size to estimate the partial trajectory resulting in anomalousness. The trajectory detection system with two-phase outliers upon trajectory data streams was improved in [263]. The two phases are the trajectory simplification and the outlier detection.

In [450], authors developed a trajectory detection method based on a Recurrent Neural Network (RNN). The authors extracted drivers' behaviors within a sliding window and used the fixed-length deep representations for the feature sequence. The authors grouped the representations into clusters before detection. CNN was first introduced in [208], which is further utilized in fields such as natural language processing and speech recognition. A CNN-based trajectory prediction method was improved in [288]. The method simplifies the network structure and utilizes the trajectory structure (spatio-temporal consistency). The experimental results show that the CNN-based trajectory prediction method can detect illegal trajectories with a high AUC score.

Our work obfuscates drivers' trajectories in the road networks to avoid the generation of off-road obfuscated locations. Then, an illegal trajectory detection scheme based on CNN is introduced in this chapter. The introduced scheme does not expose the drivers' actual trajectories and achieves high detection accuracy. The introduced scheme can detect illegal trajectories even if the drivers use various privacy parameters to obfuscate the locations. To the best of our knowledge, we are the first work that detects illegal locations in real road networks, which is almost the actual usage scenario of a smart city.

4.3 Obfuscation and Detection Scheme

In this chapter, we introduce an adaptive obfuscation scheme to protect location privacy in road networks. Then, we introduce an illegal trajectory detection system with CNN to identify the legal and illegal trajectory from the obfuscated trajectory. In our model, X_m is the m-th trajectory consists of the location points sequence $(x_{m1}, x_{m2}, \cdots, x_{mn})$. $x_{mi} = (lat_i, lon_i)$ is a tuple that stands for the coordinate (i.e., latitude lat_i and longitude lon_i) of a location.

4.3.1 System Model

The existing obfuscation studies [386, 12, 496] pay attention to protecting the drivers' location data in a 2D map, which generates off-road locations (e.g., railroads and rivers). Adversaries can exclude the off-road obfuscated locations from real trajectory data when they identify the obfuscated locations that are off-road. The Euclidean distance between the obtained off-road location and the nearby road can be utilized by the adversaries to estimate the actual location. In this chapter, we control the obfuscated candidates to avoid the off-road data and guarantee that all obfuscated locations are on-road.

4.3.2 Dynamic Obfuscation Scheme

Definition 4.1 ***Geo-Indistinguishability*** *[32]: Let P be a probabilistic function. Let X and Z be a set of the actual location candidates and obfuscated locations candidates, respectively. The K represents the mechanism that uses the probability $P(Z)$ to map an element* *in X to an element in Z. K is ϵ-geo-distinguishable, if and only if for all x, x' has:*

$$d_{\mathcal{P}}\left(K(x), K\left(x'\right)\right) \leq \epsilon d\left(x, x'\right) \tag{4.1}$$

The $X_m = \{x_{m1}, \ldots, x_{mn}\}$ is the raw path that indicates the actual trajectories, while $Z_m = \{z_{m1}, \ldots, z_{mn}\}$ indicates the obfuscated trajectories. We utilize 2D Laplace noise $D_\epsilon(x)(z) = \frac{\epsilon^2}{2\pi}e^{-\epsilon d(x,z)}$ as obfuscation distribution in this chapter. The reason is that the 2D Laplace noise ensures that z_{mi} is distributed around x_{mi}. The probability of z_{mi} with 2D Laplace noise decreases with the increasing of $d(x_{mi}, z_{mi})$ ($d(\cdot,\cdot)$ is the Euclidean distance in this chapter). Theϵ-geo-distinguishable privacy condition is also satisfied with the 2D Laplace noise.

4.3.3 Adaptive Location Privacy-Preserving Scheme

A new adaptive location privacy-preserving scheme is introduced in this chapter. The introduced scheme sets ϵ correlated to the generated obfuscated locations. By utilizing the introduced adaptive location privacy-preserving scheme, we boost the randomness of the noise generation, which increases the difficulty of inferring the actual location of a driver.

We configure ϵ into the high, medium, and low-privacy levels. We set the average distance of the obfuscated location to r when there is a low level ϵ. A medium and a high level ϵ have an average distance of $1.5r$ and $2.25r$, respectively. The introduced scheme obfuscates every single location point in each continuous trajectory. As the starting location and destination are more sensitive to a driver, we set the highest privacy level for the two locations. For other locations x_i in the trajectory, the obfuscation parameters are related to the Euclidean distance $d(x_i, z_{i-1})$. We set two different thresholds $D1$ and $D2$ to divide $d(x_i, z_{i-1})$ into three types, where $D1 \neq D2$. When the value of $d(x_i, z_{i-1})$ is bigger than the values of $D1$ and $D2$, the correlation between x_i and z_{i-1} is weak. In this case, a low-level noise is added to the actual location data. When the value of $d(x_i, z_{i-1})$ is less than the values of $D1$ and $D2$, x_i and z_{i-1} is close in road networks (i.e., high correlation). Therefore, we add noise with a high-privacy level when obfuscating actual locations in this scenario. Otherwise, we set ϵ as a medium privacy level in the introduced obfuscation scheme.

The introduced scheme reduces the correlation between x_i and z_{i-1}. Hence, the adversary cannot infer ϵ by analyzing the prior knowledge and the obtained obfuscated locations within a specific time window. The adversary cannot predict the driver's future locations because the value of ϵ is changing, which increases the difficulty of attacks.

The selection of ϵ in the introduced scheme aims to balance location privacy and data utility. A large amount of noise is required to achieve a high-privacy level but leads to a low QoS of LBS. The different ϵ can provide customized geographic location accuracy, which is suitable for various LBS requirements. For example, location-sensitive LBS (e.g., navigation) needs high-accuracy location data, so the scheme ought to utilize a high ϵ to provide a high data utility. For location-insensitive LBS, e.g., weather forecasts, the scheme can employ a low ϵ for a high-privacy level.

The amount of added noise is controlled in the introduced scheme to balance data utility and location privacy. We use z within a region based on x to set the upper bound of the QoS loss and that of the capability of location privacy, i.e., $d_{max}(x, z)$. If the distance $d(x, z)$ between the actual location x and the obfuscated location z exceeds $d_{max}(x, z)$, the introduced scheme will obfuscate the driver's actual location again.

The introduced scheme generates obfuscated locations $z_{m1}, \ldots, z_n$ and maps the obfuscated locations to the nearest road. Therefore, the introduced scheme obfuscates actual trajectories to reachable on-road locations.

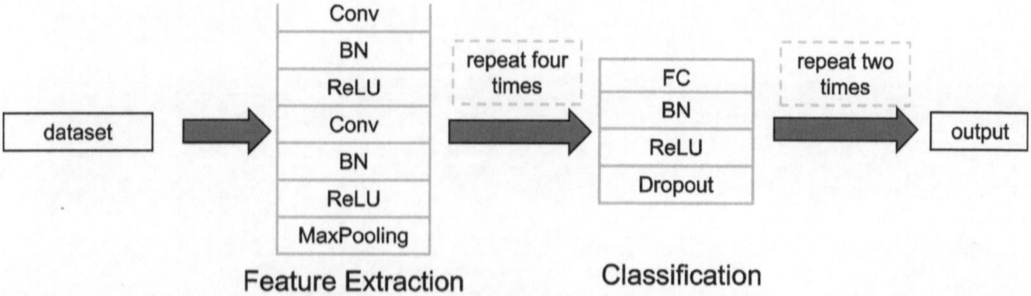

FIGURE 4.1: Our CNN model architecture.

4.3.4 Illegal Trajectories Detection Based on CNN

A Two-Dimensional Convolutional Neural Network (2D-CNN) model is applied in the introduced scheme to detect illegal trajectories. The introduced model is shown in Figure 4.1.

One-dimensional CNN (1D-CNN) consists of the convolutional layer, the subsampling layer, and the optional fully-connected layers. In convolutional neural networks, the convolutional layer is the central part that analyzes the input data to extract classification features. The Separate feature extractor contains multiple convolution kernels. The convolutional layer of the CNN model can extract the Spatial-temporal correlation of the trajectory. After extracting features in the convolutional layer, the pooling step starts. The weight parameter redundancy is solved by the local connection and weight sharing. However, the over-fitting problems arise due to the CNN model degrading the generalization performance. With the extracted features from the convolution layer and pooling, the CNN model can reduce the data dimension while retaining the value of the principal feature map.

Our work improves the architecture of 1D-CNN with the 2D-CNN model, which is widely utilized in the classification of images (e.g., ResNet [156], VGG [365], and GoogleNet [349]). The introduced model is described as follows.

- In advance of the maximum pooling layer, two convolutional layers are employed in our model. Thus, the introduced model can extract features effectively.
- We add a normalization (BN) layer after the two convolutional layers to retain the spatial-temporal correlation of locations. The Fully Connected (FC) layers are combined with the dropout layers and BN in the introduced model to avoid the FC layers leading to over-fitting issues.

4.4 Evaluation

We employ real-world road network information of Porto, which is extracted from OpenStreetMap (OSM) to evaluate the introduced scheme. OSM is popular in LBS applications, such as route planning and the geocoding of address [259]. The extracted road network is shown in Figure 4.2. The Portugal taxi trajectory dataset,[2] whose recorded location data has 15 15-second time intervals to build the trajectory, is used in our experiment.

[2]Portugal taxi trajectory dataset[Online]. Available: https://www.kaggle.com/c/pkdd-15-predict-taxi-service-trajectory-i

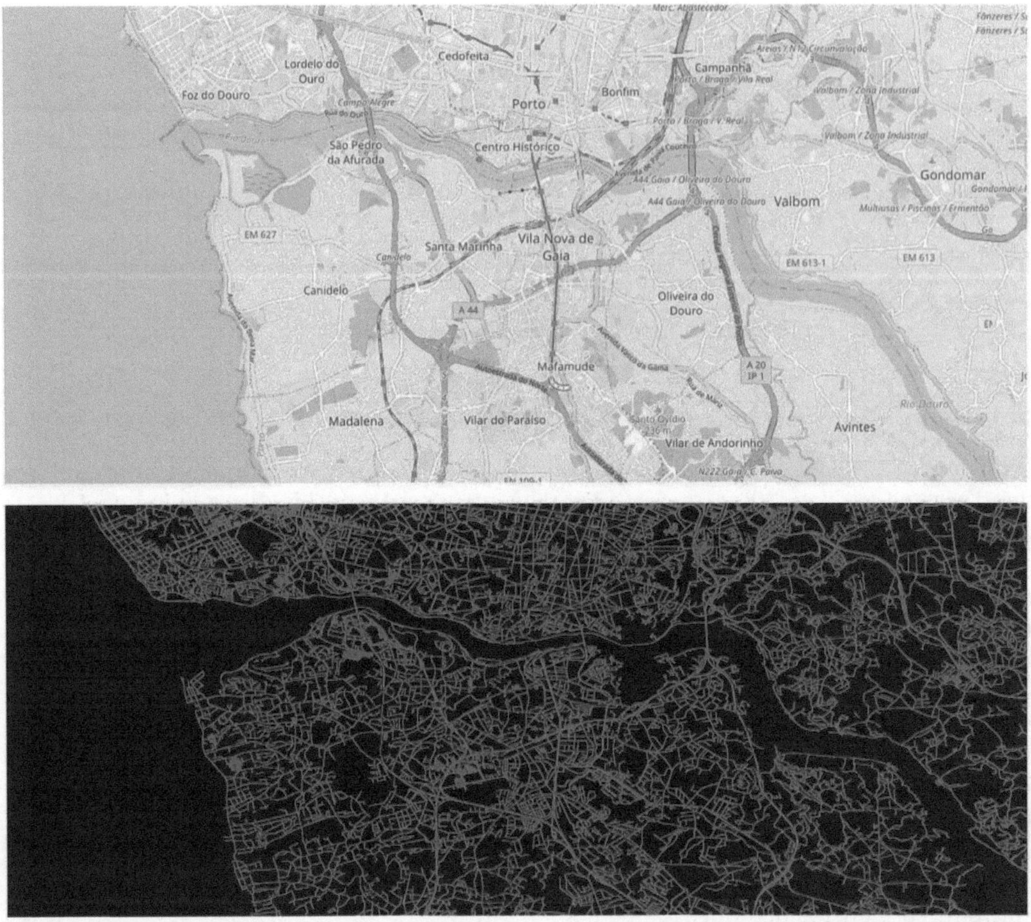

FIGURE 4.2: Generation of the road network. Upper: real road map in OSM; Bottom: the generated road network.

4.4.1 Illegal Trajectories

As far as we know, no public datasets are labeled with illegal and legal trajectories. Three methods are popular for generating illegal trajectories according to legal trajectories.

1. Insertion trajectory from other sources of trajectory dataset as the illegal trajectories [292].
2. Division of the trajectories data into legal and illegal datasets [138].
3. Combination of the above two methods [386].

In this chapter, we generate illegal trajectories by utilizing legal trajectories. The generated illegal trajectories are employed in our classification experiments. Hence, legal and illegal trajectories in the training and detection come from the same dataset to reduce deviation.

4.4.2 Simulation Results

We evaluate the detection capability of the introduced scheme with the generated trajectory data and the public trajectory dataset of the Portugal taxi. In the experiment, we configure ϵ to control the level of Laplace noise.

4.4.2.1 Experimental Setting

We use the public real-world trajectory dataset and the synthetic data as follows.

Real-World Trajectory: We employ the public trajectory dataset of Portugal taxi, which has approximately 1.7 million trajectory data.

Synthetic Data: We extract half of the trajectory data from the Portugal taxi dataset to generate the illegal data. In this chapter, the illegal trajectories considered have two forms, speed anomaly and path anomaly. The two forms represent malicious actions, i.e., speeding and detour, respectively. The illegal trajectories are generated as follows:

- **Speed Anomaly:** The maximum speeds of vehicles are limited in road networks. Malicious drivers drive faster than the limitations to obtain more profit within a period. The utilized trajectory dataset contains a timestamp; we delete x_i in the selected continuous trajectory X with a certain probability and reassign the timestamp. The trajectory X has a higher speed than the speed limitation.

- **Path Anomaly:** Malicious drivers can also select a longer route than the recommended route, i.e., path anomaly. We utilize multiple legal trajectories (e.g., three trajectories X_a, X_b, X_c) to generate path anomaly illegal trajectories. The starting location and destination of X_a are denoted as x_{a1} and x_{an}, respectively, at shown in Figure 4.3. We employ the trajectories X_b and X_c to intersect[3] the trajectory X_a at location x_{ai}, x_{aj}. We also combine the trajectories X_b and X_c at location x_{bl}. Then, we obtain an illegal trajectory, which is a path anomaly. The starting location and destination of the generated trajectory are x_{a1} and x_{an}, respectively, but the route distance between x_{a1} and x_{an} is longer than it should be. In this experiment, the length of the generated trajectories is set to be at least 1.6 times that of the legal trajectories.

The above types of illegal trajectories include the most common categories of malicious activities in road networks. We use more than 600,000 legal and illegal trajectories to evaluate our introduced detection scheme.

4.4.2.2 Implementation

The configurations of the introduced mechanism that is implemented in our experiments are as follows.

We use $\epsilon_0 = 0$ to stand for the non-protected situation. We configure the obfuscation radius as the average Euclidean distance between x_i and z_i when using noise level ϵ_i. When $\epsilon_0 = 0$, the average obfuscation radius is 0 m. We set ϵ_1 and ϵ_2 with obfuscation radii 100 m and 1000 m, respectively.

4.4.2.3 Experimental Results

We use Python to conduct the experiments. After running the experiment five times, we take the average value of the results.

A large amount of noise decreases the accuracy rate of the CNN model, as shown in Table 4.1. We compare the Receiver Operating characteristics Curve (ROC) with various settings of ϵ, as shown in Figure 4.4. When applying ϵ_1 and ϵ_2, the introduced scheme achieves a high-value accuracy rate of 0.94 on the AUC score. Compared with that of ϵ_1, the AUC score of ϵ_2 has been reduced by 0.14 to 0.80.

[3] The intersection stands for the points of the two trajectories whose distance are within a certain range.

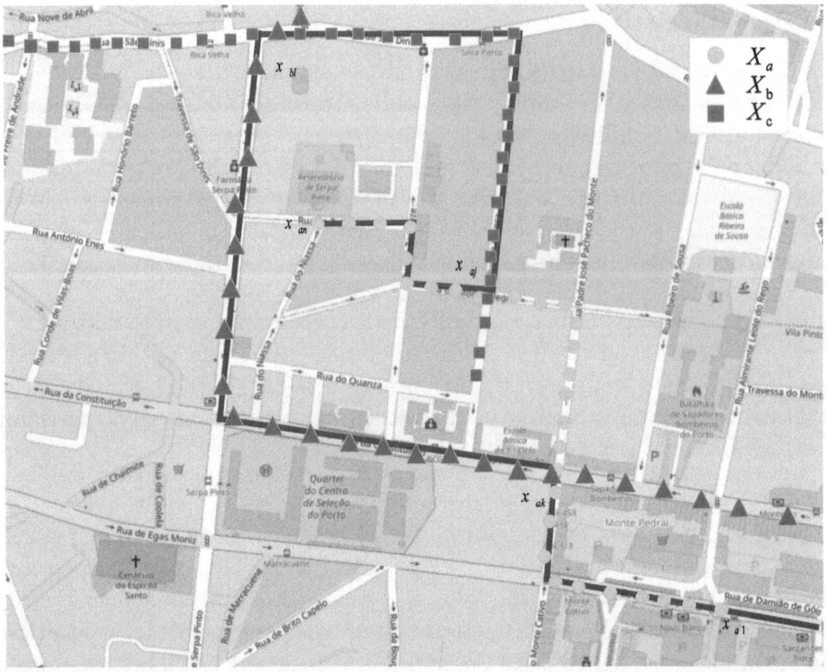

FIGURE 4.3: Example of an anomaly generated path.

4.4.3 Contrast

We use the same trajectory dataset as the existing related work [386]. When detecting illegal trajectories, the introduced scheme achieves a higher accuracy rate than the scheme developed in [386]. The introduced scheme considers a more complex environment, i.e., road networks, than the 2D plane environment considered in [386] The introduced scheme utilizes a different obfuscation process and recognition dataset, so we do not compare the accuracy of the two schemes. Compared with the schemes in [386], the introduced scheme has advantages as follows.

- The scheme in [386] employs fixed parameters in the obfuscation process, while the introduced scheme dynamically calculates the parameters. The dynamically calculated parameters provide higher privacy protection capability and high data utility than the scheme in [386].
- The scheme in [386] manually injects trajectories to generate illegal trajectories. The illegal trajectories generated in this chapter are closer to the real world than that of the scheme in [386]. The generated illegal trajectories in this chapter are indistinguishable from the real trajectories, increasing the difficulty of detecting illegal trajectories. Under the strict assumption, the introduced scheme still achieves a higher accuracy rate than the existing work [386].

TABLE 4.1: Experimental accuracy.

ϵ	Average Value of the Noise Radius (m)	Accuracy
ϵ_0	0	93.1%
ϵ_1	100	86.1%
ϵ_2	1,000	72.5%

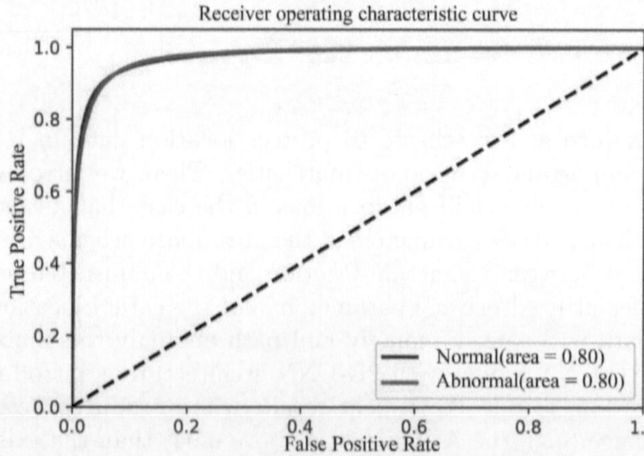

(a) ϵ takes 0 m, i.e., ϵ equals to ϵ_0

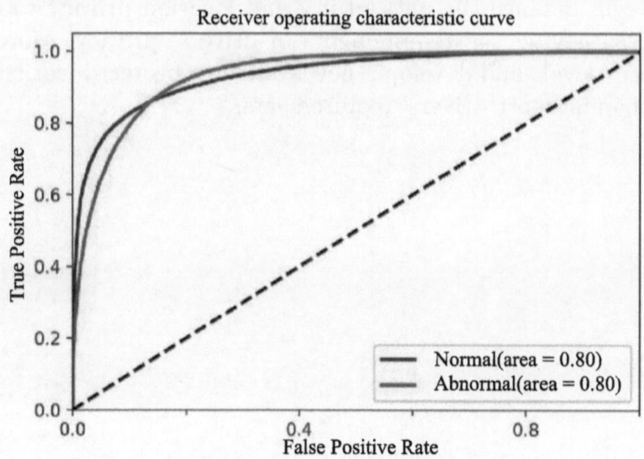

(b) ϵ takes 100 m, i.e., ϵ equals to ϵ_1

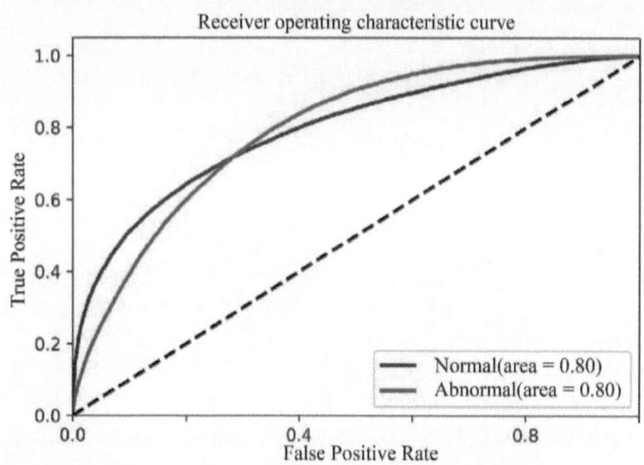

(c) ϵ takes 1,000 m, i.e., ϵ equals to ϵ_2

FIGURE 4.4: ROC under three different parameters.

- We employ the real-world dataset and road networks to evaluate the introduced scheme. Thus, the introduced scheme in this chapter has practical meaning.

4.5 Conclusion

In this chapter, we introduce a new scheme to protect location data in real-world road networks adaptively, an important scenario of smart cities. Then, we introduced an illegal trajectory detection scheme to detect illegal locations in the case that all drivers are protected in road networks. The privacy parameters of the introduced scheme are calculated by considering the correlation between the actual location and the obfuscated location. Thus, the adversary cannot infer utilized privacy parameters and the actual locations. We generated illegal trajectory data with speed anomaly and path anomaly to simulate real-world malicious driving. The 1D-CNN model with 2D-CNN architecture is introduced to detect illegal trajectories. According to our experiment results, the introduced detection scheme achieves better performance (e.g., the AUC score is above 0.94) than the existing works in road networks.

In future work, we will balance the data utility and location privacy to maximize the data availability while satisfying the requirements of drivers' privacy. Moreover, we will assess the drivers' privacy levels and develop a new system to protect location data privacy with the capability to handle most drivers' requirements.

5

Cloaking Region Obfuscation for RN-I and Location Privacy

Using RN-Indistinguishability, we design the Cloaking Region Obfuscation (CRO) mechanism that employs the route distances to quantify the indistinguishability of road locations. We prove that the CRO mechanism satisfies the RN-Indistinguishability. The CRO mechanism can be extended with general road network features without breaching differential privacy.

Section 5.1 provides the background of location and identity privacy protection in vehicular networks. Section 5.2 studies the existing works of location and identity protection. Section 5.3 illustrates the system and adversary model. Section 5.4 shows the introduced CRO mechanism, which is evaluated in Section 5.5. Section 5.6 gives a conclusion of the introduced CRO mechanism.

5.1 Introduction

Location-based services (LBS) provide vehicular applications access to the location data [175]. LBS servers are able to obtain drivers' geographical locations, which puts drivers' location privacy at risk [191]. Adversaries can infer drivers' information from the location data after gaining unlawful access to the LBS servers [296]. Adversaries can also attack and eavesdrop on the communications between the LBS servers and drivers to obtain drivers' location data [453, 485]. Unlike the general mobile LBS users, vehicles are more accessible to track due to the fact that the vehicles' trajectories are considerably predictable in road networks [285].

Obfuscation-based mechanisms have been developed to protect the LBS users' location privacy against eavesdroppers and malicious LBS servers [219, 388, 31]. Users can perturb their real locations locally by exploiting differential privacy mechanisms that add controllable noises to the location data [482]. The magnitudes of the additive noises under an indistinguishability constraint are determined based on the distance between an actual location and its obfuscated version [31]. Drivers can use the obfuscated location in their requests to an LBS server [321]. Although the obfuscation mechanisms can penalize location accuracy, they can still be used in location-insensitive LBS, such as location-based recommendations [271].

Vehicles in road networks have not been well considered in existing obfuscation-based mechanisms. General-purpose obfuscation-based mechanisms (e.g., two-dimensional (2D) Laplace location privacy-preserving mechanisms) use the Euclidean distance to measure the distance between an actual location and the corresponding obfuscated location [31, 166, 388]. However, the mechanisms underestimate the distance in road networks due to the fact that

DOI: 10.1201/9781003518044-5

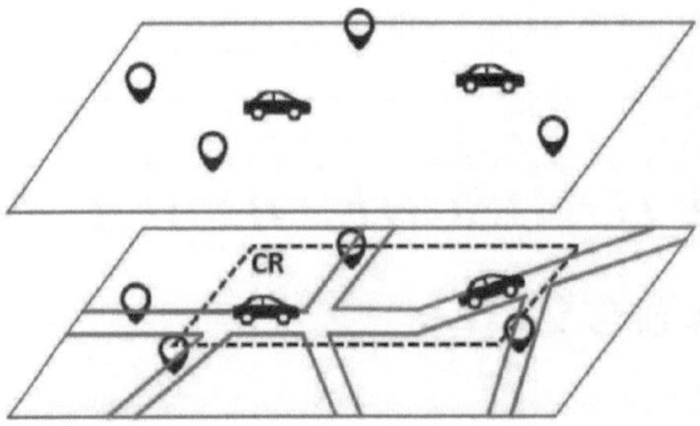

FIGURE 5.1: The difference between the existing Geo-I mechanism and the introduced Cloaking Region Obfuscation (CRO) mechanism.

the Euclidean distance is less than, or equal to, the route distance between two locations [33]. Another limitation of the existing obfuscation-based mechanisms is that the mechanisms may generate off-road locations, e.g., locations in a river [381]. Extra effort (at the cost of high computational complexity) is needed to avoid those off-road locations [184]. In existing location obfuscation mechanisms [31, 166, 388], vehicles independently obfuscate their locations, which increases the risk of location data leakage [403].

In this chapter, we define a new notion of Road Network-Indistinguishability (RN-Indistinguishability), where the indistinguishability between locations is measured with road network features (e.g., road network density, speed limits, and route distance). The difference between the introduced RN-Indistinguishability and the existing Geo-I is shown in Figure 5.1. Based on RN-Indistinguishability, we introduce a new Cloaking Region Obfuscation (CRO) mechanism that protects location privacy in road networks by considering road network features (e.g., route distance). Vehicles are obfuscated to other regions, according to the indistinguishability between the obfuscated region and the actual cloaking region. The CRO allows multiple vehicles to cooperate. Vehicles in the same cloaking region are uniformly obfuscated such that they are indistinguishable from each other and can cooperate to improve privacy protection.

The topics of this chapter are summarized as follows,

1) We develop differential privacy in road networks and introduce the RN-Indistinguishability that quantifies road network location privacy with unique road network features.

2) By using the RN-Indistinguishability, we design the CRO mechanism that employs the route distances to quantify the indistinguishability of locations on roads. We prove that the CRO mechanism satisfies the RN-Indistinguishability. The CRO mechanism can be extended with general road network features without breaching differential privacy.

We evaluate the CRO mechanism with real-world road networks in comparison with the state-of-the-art 2D Laplace location privacy-preserving mechanisms [166, 12]. Experiment results confirm the superiority of the CRO mechanism in terms of privacy-preserving level and data utility.

5.2 Related Work

Location privacy-preserving mechanisms can be classified into user-side mechanisms, server-side mechanisms, and channel-side mechanisms [183]. Studies mainly use obfuscation,

TABLE 5.1: Summary of notation and abbreviations.

Notation	Description
$\mathcal{G}$	The graph transformed from a road network
l	The location of a cloaking region
$d_{\mathcal{G}}(l, l')$	The shortest route distance between l and l'
$\mathcal{D}_{\mathcal{G}}(l, l')$	The indistinguishability between l and l'
ϵ	Privacy budget
$\mathbb{R}^2$	Obfuscation area
r	The radius of obfuscation areas
SD	Shift Distance
AEE	Adversary Estimation Error
CRO	Cloaking Region Obfuscation

anonymity, and cryptography to realize these three mechanisms [196]. Obfuscation mechanisms aim to reduce the precision of location data by adding noise in the actual location data [55]. Obfuscation mechanisms suit the LBSs that perform fine with coarse location data [191]. Anonymity mechanisms allow vehicles to use pseudonyms to hide vehicles' real identities [191]. However, the spatial-temporal correlation of the vehicles' trajectories can be estimated by adversaries [315]. Cryptographic algorithms can be used to encrypt the actual location data. Alternatively, certified credentials can be allocated to drivers from trusted authorities [37]. The cryptographic mechanisms have a limitation in time-sensitive LBSs due to the potentially high computational complexity and untrusted entities [360].

Differential Privacy (DP) technology is widely used in obfuscation mechanisms [426]. It protects individuals' private information indistinguishable when publishing aggregated data [415, 134].

Most of the existing obfuscation mechanisms protect location privacy on two-dimensional (2D) maps. Andrés *et al.* [31] developed Geo-Indistinguishability (Geo-I) based on differential privacy. The authors of [200] employed the Laplace mechanism to realize Geo-Indistinguishability on a 2D plane. The Geo-I reports obfuscated locations surrounding the actual location based on the obfuscated probability distribution when a driver requests LBS. Hua *et al.* [166] introduced additional servers to assist Geo-I obfuscation process. The authors divided the map into several areas by introducing extra servers. A driver employs the actual location of the servers to calculate the obfuscated locations. Rather than dividing maps, the mechanism developed in [321] splits roads into same-length road sections. The authors employed the route distance between two road sections to measure the indistinguishability of the Geo-Indistinguishability. Drivers who are located in the same road section are obfuscated in the same rule. The Graph-Exponential Mechanism (GEM) evaluates the privacy-preserving capability of the Geo-Indistinguishability in road networks [388]. The GEM sets the connections (such as turns, intersections, bifurcates, and joins) in a road network as the obfuscation candidates. The mechanism maps drivers' actual locations to connections and obfuscates connections based on Geo-Indistinguishability.

The cloaking region, in which drivers' private data are uniformly processed, is introduced to spatially or temporally diminish the accuracy of the drivers' actual location data in LBS queries [175]. The existing cloaking-region mechanisms focus on anonymity and obfuscation in a 2D map rather than in road networks. Luo *et al.* [251] improved a distributed k-anonymity (spatially) mechanism by using blockchain. The authors improved the construction process of the anonymous cloaking region based on the historical information recorded on publicly available blocks. Yang *et al.* [448] designed a reputation calculation algorithm based on blockchain. The authors used trusted nodes to create cloaking regions, in which the location data can be protected. To curb the malicious behaviors, the authors improved a punishment factor in their algorithm. Rather than blockchain, Li *et al.* [212] protected location privacy with socially intimate fogs. The authors transferred many LBS requests to social intimate fogs, which improves efficiency and increases communication delay (temporally). The authors improved a fog-based LBS agent that can re-encrypt multiple location data before transmitting it to LBS. Cloaking region establishment is a reputation-computation question in the anonymity area (e.g., [251] and [448]), which is not the key point of this chapter.

To the best of our knowledge, none of the Geo-I and existing cloaking-region mechanisms obfuscate actual locations by considering the road network features, such as route distance and road network topology.

5.3 System Model

This section illustrates the considered road network model and adversary model, which are detailed as follows.

5.3.1 Road Network Model

Drivers request various LBS when driving in road networks following traffic rules. Considering the untrustworthiness of the LBS servers, the drivers do not want to share their precise location data with the servers. The drivers can protect their location privacy by employing obfuscation mechanisms and reporting obfuscated locations to the servers.

5.3.1.1 Road Network

A road network is modeled as a directed and weighted graph $\mathcal{G}$. The road network nodes are the connections (e.g., turns, intersections, bifurcates, and joints) mapped to the same geo-locations in the real world. The weight of each edge indicates the length of the corresponding road section between the two connections.

5.3.1.2 Cloaking Regions (CRs)

We evenly divide road networks into CRs of the same area size. A CR is empty if there is no connection or road inside it. The empty CR has been removed from our model.

5.3.1.3 Actual Location l

In each CR, we select the closest on-road point to the center of the CR to represent the CR. The location of the n-th CR, denoted by CR_n, is represented by the location of the selected point in CR_n, denoted by, l_n. The locations of vehicles in CR_n use l_n as their actual locations.

5.3.1.4 Obfuscation Area $\mathbb{R}^2$

The obfuscation area $\mathbb{R}^2$ of l is a circular region with radius r and centered on l, in which the locations of the vehicles can be obfuscated based on the obfuscation probability. A CR_n is inside $\mathbb{R}^2$ if its center location l_n is in $\mathbb{R}^2$. All CRs in $\mathbb{R}^2$ are obfuscation candidates of the CR with the location l.

5.3.1.5 Route Distance

The route distance between a pair of CRs (CR_i and CR_j) is measured using the route distance between l_i and l_j, i.e., $d_{\mathcal{G}}(l_i, l_i)$.

5.3.1.6 LBS Request

The LBS requests are sent from vehicles to the LBS servers. In our model, an unprotected LBS query q from a vehicle in a CR is as follows,

$$q = (l, I),$$

where l is the location of the CR within which the vehicle is. I is the other content of the query.

We assume that the LBS servers are not trusted and may disclose the drivers' data to adversaries. We use differential privacy to protect drivers' location privacy. When a driver's location needs to be protected, the driver can access the LBS server by sending a private query q' with an obfuscated location, as given by

$$q' = (l', I), \tag{5.1}$$

where l' is the location of the obfuscated cloaking region selected based on l.

We define shift distance (SD) to measure the data utility in the queries affected by the obfuscation mechanism. Given a set of queries from CR with l, the average SD of the cloaking region ($SD(l)$) is given by [178, eq. 3]

$$SD(l) = \frac{1}{K}\sum_{k=1}^{K} d_{\mathcal{G}}(l, l'_k), \tag{5.2}$$

where K is the number of queries from the drivers in the CR. l'_k is the location of the obfuscated CR in the k-th query. The SD measures the data utility of obfuscated locations. A smaller SD indicates a higher data utility and a potentially better LBS. The data utility decreases when the drivers employ obfuscation mechanisms to protect their location information.

5.3.2 Adversary Model

Both the external and internal passive attackers are considered in this chapter, as shown in Figure 5.2. External adversaries can eavesdrop on the LBS messages exchanged between the LBS servers and the vehicles to obtain the vehicles' locations. Internal adversaries can collude with the LBS servers to obtain the vehicles' queries. We assume that the adversaries are fully aware of the drivers' prior trajectories and the obfuscation mechanism [321].

We follow the adversary model in [433]. The actual location of a driver is unknown, and no other parties (e.g., adversaries and servers) can directly observe the driver's actual location. The existing anonymity mechanisms can protect the drivers' identities so that adversaries cannot track drivers' trajectories by analyzing identity information. Other information, such as obfuscation mechanisms and parameters, is known to all parties.

Obfuscation Probability: Given the location l of the CR within which vehicle in, location obfuscation mechanisms select an obfuscated CR with location l' based on the obfuscation probability $\Pr[l'|l]$.

The adversaries can infer the probability that the obfuscated location l' is generated by the location l in $\mathcal{G}$ is given by [433, eq. 1]

$$\Pr[l_i|l'] = \frac{\Pr[l'|l_i]\Pr[l_i]}{\sum_{l_j \in \mathbb{R}^2} \Pr[l'|l_j]\Pr[l_j]}, \tag{5.3}$$

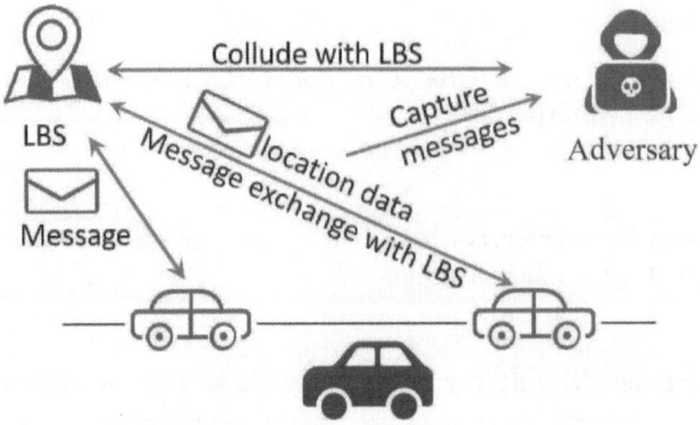

FIGURE 5.2: Adversary model.

where $\Pr[l_i]$ and $\Pr[l_j]$ are the prior probability that the driver locates at the CR_i (with location l_i) and CR_j (with location l_j) according to the prior trajectories, respectively. The prior probability is based on the driver's historical trajectory. $\Pr[l'|l_j]$ is the probability of the region CR' being selected as the obfuscated version of the actual region CR for a vehicle.

The adversaries select the CR, $\hat{CR}$ (with location $\hat{l}$), as the actual cloaking region of the driver, if $\hat{l}$ has

$$\Pr[\hat{l}|l'] = \max_{l_j \in \mathcal{G}} \Pr[l_j|l']. \tag{5.4}$$

The shortest route distance between the two cloaking regions CR (with location l) and $\hat{CR}$ (with location $\hat{l}$) measures the Adversary Estimation Error (AEE), which evaluates the protection effect of an obfuscation mechanism on location privacy.

As illustrated in Figure 5.3, the AEE is $d_{\mathcal{G}}(l, \hat{l})$ and the SD is $d_{\mathcal{G}}(l, l')$. Given a set of queries from CR with l, $\hat{CR}_k$ with $\hat{l_k}$ is the k-th derived CR, the average AEE of the CR ($AEE(l)$) is given by

$$AEE(l) = \frac{1}{K} \sum_{k=1}^{K} d_{\mathcal{G}}(l, \hat{l_k}). \tag{5.5}$$

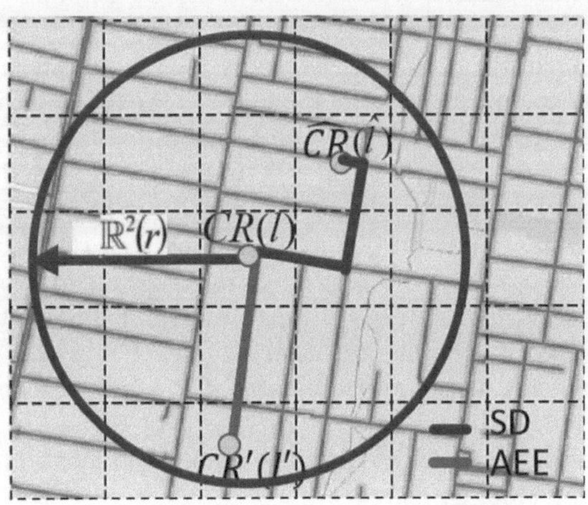

FIGURE 5.3: An example of the introduced CRO algorithm.

Figure 5.3 provides a visual representation of the introduced CRO algorithm. It shows the cloaking region CR and another cloaking region $\hat{CR}$. he adversary's estimation error is represented by the shortest path $d_{\mathcal{G}}(l,\hat{l})$ between l and $\hat{l}$. The solid black line represents the shortest distance (SD) between the actual location l and a derived location l' within the cloaking region. The line segments indicate the potential paths used to calculate the AEE, showing the variations in the estimated location based on different queries.

5.4 Cloaking Region Obfuscation Mechanism

In this section, we define RN-Indistinguishability with general road network features (e.g., route distances, speed limits, and road network densities) based on the definition of Geo-I [31]. The RN-Indistinguishability employs road network features to measure the indistinguishability of locations in road networks. By using the route distance as the metric of RN-Indistinguishability, we introduce the Cloaking Region Obfuscation (CRO) mechanism to protect drivers' location privacy in road networks. The introduced CRO mechanism ensures that adversaries cannot infer the drivers' actual locations based on the obfuscated location data. We prove that the CRO mechanism satisfies the RN-Indistinguishability and can be generalized with multiple road network features.

5.4.1 Road Network-Indistinguishability

The Geo-I mechanism can protect the location privacy in 2D Euclidean spaces [31]. The Geo-I mechanism makes the actual locations Geo-indistinguishable from any adversaries by using differential privacy to perturb the drivers' actual locations. In contrast, the RN-Indistinguishability assesses on-road privacy-preserving mechanisms and is defined as follows.

Definition 5.1 (RN-Indistinguishability) *A mechanism satisfies $(\epsilon,\mathbb{R}^2)$-RN-Indistinguishability if and only if, with an obfuscated location l', l' in $\mathbb{R}^2$), any location pair $(l_i,\, l_j)$ in $\mathbb{R}^2$ have*

$$\frac{\Pr[l_i|l']}{\Pr[l_j|l']} \le e^{\epsilon \mathcal{D}_{\mathcal{G}}(l_i,l_j)} \frac{\Pr[l_i]}{\Pr[l_j]}, \tag{5.6}$$

where $\mathcal{D}_{\mathcal{G}}(l_i,l_j)$ is the indistinguishability between l_i and l_j in road networks. ϵ is the privacy budget. $\Pr[l_i|l']$ and $\Pr[l_j|l']$ are the probabilities that l' is obfuscated based on l_i and l_j, respectively.

In this chapter, RN-Indistinguishability is employed to obfuscate cloaking regions. With the cloaking region CR_i and CR_j, $\mathcal{D}_{\mathcal{G}}(l_i,l_j)$ can be calculated by [224]

$$\begin{cases} \mathcal{D}_{\mathcal{G}}(l_i,l_j) = \alpha_1 \dfrac{d_{\mathcal{G}}(l_i,l_j)}{\max_{\mathbb{R}^2} d_{\mathcal{G}}} + \alpha_2 \dfrac{v_{\mathcal{G}}(l_i,l_j)}{\max_{\mathbb{R}^2} v_{\mathcal{G}}} + \alpha_3 \dfrac{den_{\mathcal{G}}(l_i,l_j)}{\max_{\mathbb{R}^2} den_{\mathcal{G}}} + \cdots; & (5.7) \\ \alpha_1 + \alpha_2 + \alpha_3 + \cdots = 1, & (5.8) \end{cases}$$

where the route distances $d_{\mathcal{G}}(l_i,l_j)$, the speed limit difference $v_{\mathcal{G}}(l_i,l_j)$, the road network density difference $den_{\mathcal{G}}(l_i,l_j)$, and other road network features can be considered. α_1, α_2, and α_3 are weighting factors. $\max_{\mathbb{R}^2} d_{\mathcal{G}}$, $\max_{\mathbb{R}^2} v_{\mathcal{G}}$, and $\max_{\mathbb{R}^2} den_{\mathcal{G}}$ are the maximum values of route distance, speed limit differences, and road network density differences in $\mathbb{R}^2$, respectively.

By substituting (5.3) into (5.6), we can have

$$\begin{aligned}
\frac{\frac{\Pr[l'|l_i]\Pr[l_i]}{\sum_{l_k\in\mathbb{R}^2}\Pr[l'|l_k]\Pr[l_k]}}{\frac{\Pr[l'|l_j]\Pr[l_j]}{\sum_{l_k\in\mathbb{R}^2}\Pr[l'|l_k]\Pr[l_k]}} &\le e^{\epsilon\mathcal{D}_{\mathcal{G}}(l_i,l_j)}\frac{\Pr[l_i]}{\Pr[l_j]};\\
\frac{\Pr[l'|l_i]\Pr[l_i]}{\Pr[l'|l_j]\Pr[l_j]} &\le e^{\epsilon\mathcal{D}_{\mathcal{G}}(l_i,l_j)}\frac{\Pr[l_i]}{\Pr[l_j]};\\
\frac{\Pr[l'|l_i]}{\Pr[l'|l_j]} &\le e^{\epsilon\mathcal{D}_{\mathcal{G}}(l_i,l_j)};\\
\Pr[l'|l_i] &\le e^{\epsilon\mathcal{D}_{\mathcal{G}}(l_i,l_j)}\Pr[l'|l_j],
\end{aligned} \tag{5.9}$$

where the two cloaking regions CR_i (with location l_i) and CR_j (with location l_j) are RN-indistinguishable to any adversaries. $\Pr[l'|l_i]$ and $\Pr[l'|l_j]$ are the probabilities of the region CR' being selected as the obfuscation version of the actual region CR_i and CR_j, respectively

5.4.2 Cloaking Region Obfuscation

In this subsection, we introduce a CRO mechanism based on the RN-Indistinguishability. We set $\alpha_1 = 1$ and the other weighting factors to be 0, which indicates that we only consider $d_{\mathcal{G}}(l_i, l_j)$ as the metric of RN-Indistinguishability.

In the CRO mechanism, we obfuscate the location of the cloaking region, where the driver actually locates, and use its location in the LBS query. All the drivers in the same cloaking region are obfuscated following the same rule and, therefore, are indistinguishable from each other. The cloaking regions in road networks can also prevent adversaries from identifying and locating vehicles.

The introduced CRO mechanism consists of the following steps:

- **Initialization:** The road network is evenly divided into cloaking regions. Vehicles in the cloaking region CR (with location l) specify the obfuscation area $\mathbb{R}^2$, centering l with radius r. Cloaking region CR_i (with location l_i) is in $\mathbb{R}^2$ if l_i is inside $\mathbb{R}^2$.

- **Obfuscation:** Given a vehicle in CR, the location of the vehicle is represented by l. The vehicle probabilistically selects an obfuscated cloaking region CR' (with location l') in $\mathbb{R}^2$. The probability $\Pr[l'|l]$ is as given by

$$\Pr[l'|l] = \frac{e^{-\frac{\epsilon}{2}\frac{d_{\mathcal{G}}(l,l')}{\max_{\mathbb{R}^2} d_{\mathcal{G}}}}}{\sum\limits_{l_i\in\mathbb{R}^2} e^{-\frac{\epsilon}{2}\frac{d_{\mathcal{G}}(l,l_i)}{\max_{\mathbb{R}^2} d_{\mathcal{G}}}}}, \tag{5.10}$$

  where l_i is the location of the i-th cloaking region in $\mathbb{R}^2$. $\max_{\mathbb{R}^2} d_{\mathcal{G}}$ is the longest route distance between two cloaking regions in $\mathbb{R}^2$. All vehicles in the same cloaking region have the same obfuscation probability.

- **Finalization:** The vehicle reports a query with l' and other information to the LBS server.

In the introduced CRO algorithm, the locations of vehicles in the same cloaking are uniformly obfuscated. Thus, the location privacy-preserving capability increases with the number of vehicles.

5.4.3 Privacy Analysis

In this subsection, we prove that the CRO algorithm with the route distance metric satisfies the RN-Indistinguishability and can protect location privacy under the adversary model in Section 5.3.2.

Theorem 5.1 *With route distance metric, the CRO mechanism satisfies the (ϵ, $\mathbb{R}^2$)-RN-Indistinguishability that the locations of any two cloaking regions in an obfuscation area $\mathbb{R}^2$ are (ϵ, $\mathbb{R}^2$)-RN-indistinguishable.*

Proof 5.1 *The cloaking region obfuscation in the CRO mechanism is formulated as* (5.10). *Let*

$$\begin{cases} f(l_0) = \sum_{l \in \mathbb{R}^2} e^{-\frac{\epsilon}{2} \frac{d_{\mathcal{G}}(l_0, l)}{\max_{\mathbb{R}^2} d_{\mathcal{G}}}}; & (5.11) \\ f(l_1) = \sum_{l \in \mathbb{R}^2} e^{-\frac{\epsilon}{2} \frac{d_{\mathcal{G}}(l_1, l)}{\max_{\mathbb{R}^2} d_{\mathcal{G}}}}. & (5.12) \end{cases}$$

Given any location pair (l_0, l_1) *in* $\mathbb{R}^2$, *we have*

$$\frac{\Pr[l'|l_0]}{\Pr[l'|l_1]} = \frac{f(l_1)}{f(l_0)} e^{\frac{\epsilon}{2}\left(\frac{d_{\mathcal{G}}(l_1, l')}{\max_{\mathbb{R}^2} d_{\mathcal{G}}} - \frac{d_{\mathcal{G}}(l_0, l')}{\max_{\mathbb{R}^2} d_{\mathcal{G}}}\right)}, \tag{5.13}$$

where l' *is the location of the obfuscated cloaking region* CR'.

Then, we prove that route distance satisfies the triangle inequality in road networks, i.e., for any three cloaking regions with locations l_a, l_b, and l_c in a road network, $d_{\mathcal{G}}(l_a, l_c) - d_{\mathcal{G}}(l_b, l_c) \leq d_{\mathcal{G}}(l_a, l_b)$ holds as follows.

If l_b *is on the shortest route between* l_a *and* l_c, *we have*

$$d_{\mathcal{G}}(l_a, l_c) - d_{\mathcal{G}}(l_b, l_c) = d_{\mathcal{G}}(l_a, l_b). \tag{5.14}$$

If l_b *is not on the shortest route between* l_a *and* l_c, *by assuming that* $d_{\mathcal{G}}(l_a, l_c) - d_{\mathcal{G}}(l_b, l_c) > d_{\mathcal{G}}(l_a, l_b)$, *we have*

$$d_{\mathcal{G}}(l_a, l_c) > d_{\mathcal{G}}(l_a, l_b) + d_{\mathcal{G}}(l_b, l_c),$$

which indicates that the route between l_a *and* l_c *is longer than the route between* l_a *and* l_c *via* l_b. *Then, the shortest route between* l_a *and* l_c *should pass* l_b. *This contradicts with the assumption of* $d_{\mathcal{G}}(l_a, l_c) - d_{\mathcal{G}}(l_b, l_c) > d_{\mathcal{G}}(l_a, l_b)$. *If* l_b *is not on the shortest route between* l_0 *and* l_c, *the following inequality holds*

$$d_{\mathcal{G}}(l_a, l_c) - d_{\mathcal{G}}(l_b, l_c) \leq d_{\mathcal{G}}(l_a, l_b). \tag{5.15}$$

Combining (5.14) *and* (5.15), *the triangle inequality in road networks is proved. With the triangle inequality in road networks, we have*

$$\begin{aligned} d_{\mathcal{G}}(l_0, l_1) = d_{\mathcal{G}}(l_1, l_0) &\geq d_{\mathcal{G}}(l_1, l') - d_{\mathcal{G}}(l_0, l'); \\ \frac{d_{\mathcal{G}}(l_0, l_1)}{\max_{\mathbb{R}^2} d_{\mathcal{G}}} = \frac{d_{\mathcal{G}}(l_1, l_0)}{\max_{\mathbb{R}^2} d_{\mathcal{G}}} &\geq \frac{d_{\mathcal{G}}(l_1, l')}{\max_{\mathbb{R}^2} d_{\mathcal{G}}} - \frac{d_{\mathcal{G}}(l_0, l')}{\max_{\mathbb{R}^2} d_{\mathcal{G}}}. \end{aligned} \tag{5.16}$$

Then,

$$\frac{\Pr[l'|l_0]}{\Pr[l'|l_1]} \leq \frac{f(l_1)}{f(l_0)} e^{\frac{\epsilon}{2} \frac{d_{\mathcal{G}}(l_0, l_1)}{\max_{\mathbb{R}^2} d_{\mathcal{G}}}}. \tag{5.17}$$

By employing the triangle inequality, we further have

$$e^{-\frac{\epsilon}{2}\frac{d_{\mathcal{G}}(l_1,l')}{\max_{\mathbb{R}^2} d_{\mathcal{G}}}} \leq e^{-\frac{\epsilon}{2}\left(\frac{d_{\mathcal{G}}(l_0,l')-d_{\mathcal{G}}(l_0,l_1)}{\max_{\mathbb{R}^2} d_{\mathcal{G}}}\right)}. \tag{5.18}$$

Therefore, for all l in $\mathbb{R}^2$, the following holds

$$\sum_{l\in\mathbb{R}^2}\left(e^{-\frac{\epsilon}{2}\frac{d_{\mathcal{G}}(l_1,l)}{\max_{\mathbb{R}^2} d_{\mathcal{G}}}} - e^{-\frac{\epsilon}{2}\left(\frac{d_{\mathcal{G}}(l_0,l)-d_{\mathcal{G}}(l_0,l_1)}{\max_{\mathbb{R}^2} d_{\mathcal{G}}}\right)}\right) \leq 0, \tag{5.19}$$

which can be rewritten as

$$\sum_{l\in\mathbb{R}^2} e^{-\frac{\epsilon}{2}\frac{d_{\mathcal{G}}(l_1,l)}{\max_{\mathbb{R}^2} d_{\mathcal{G}}}} - e^{\frac{\epsilon}{2}\frac{d_{\mathcal{G}}(l_0,l_1)}{\max_{\mathbb{R}^2} d_{\mathcal{G}}}} \sum_{l\in\mathbb{R}^2} e^{-\frac{\epsilon}{2}\frac{d_{\mathcal{G}}(l_0,l)}{\max_{\mathbb{R}^2} d_{\mathcal{G}}}} \leq 0. \tag{5.20}$$

Based on the definition of $f(l_0)$, we have

$$f(l_1) - e^{\frac{\epsilon}{2}\frac{d_{\mathcal{G}}(l_0,l_1)}{\max_{\mathbb{R}^2} d_{\mathcal{G}}}} f(l_0) \leq 0, \tag{5.21}$$

which leads to

$$\frac{f(l_1)}{f(l_0)} \leq e^{\frac{\epsilon}{2}\frac{d_{\mathcal{G}}(l_0,l_1)}{\max_{\mathbb{R}^2} d_{\mathcal{G}}}}. \tag{5.22}$$

Combining (5.17) *and* (5.22), *we have*

$$\frac{\Pr[l'|l_0]}{\Pr[l'|l_1]} \leq e^{\frac{\epsilon}{2}\frac{d_{\mathcal{G}}(l_0,l_1)}{\max_{\mathbb{R}^2} d_{\mathcal{G}}}} e^{\frac{\epsilon}{2}\frac{d_{\mathcal{G}}(l_0,l_1)}{\max_{\mathbb{R}^2} d_{\mathcal{G}}}} = e^{\epsilon\frac{d_{\mathcal{G}}(l_0,l_1)}{\max_{\mathbb{R}^2} d_{\mathcal{G}}}}, \tag{5.23}$$

which satisfies (5.9). *In other words, CR_0 and CR_1 in $\mathbb{R}^2$ are $(\epsilon,\mathbb{R}^2)$-RN-indistinguishable.*

5.4.4 Generalization with Road Network Features

We have used the route distance as the only metric of the RN-Indistinguishability in the introduced CRO mechanism. In this subsection, we prove the CRO mechanism satisfies RN-Indistinguishability, when it is generalized with other road network features (i.e., route distances, speed limits, and road network densities).

Theorem 5.2 *The CRO mechanism satisfies the $(\epsilon,\mathbb{R}^2)$-RN-Indistinguishability when the employed road network features satisfy the triangle inequality.*

Proof 5.2 *With multiple road network features, the CRO mechanism satisfies $(\epsilon,\mathbb{R}^2)$-RN-Indistinguishability if*

$$\frac{\Pr[l'|l_0]}{\Pr[l'|l_1]} \leq e^{\epsilon\mathcal{D}_{\mathcal{G}}(l_0.l_1)}. \tag{5.24}$$

As shown in (5.13), *the left-hand side of* (5.24) *can be rewritten as*

$$\frac{\Pr[l'|l_0]}{\Pr[l'|l_1]} = \frac{\sum\limits_{l_j\in\mathbb{R}^2} e^{\frac{\epsilon}{2}\mathcal{D}_{\mathcal{G}}(l_j,l')}}{\sum\limits_{l_i\in\mathbb{R}^2} e^{\frac{\epsilon}{2}\mathcal{D}_{\mathcal{G}}(l_i,l')}} e^{\frac{\epsilon}{2}(\mathcal{D}_{\mathcal{G}}(l_0,l')-\mathcal{D}_{\mathcal{G}}(l_1,l'))}, \tag{5.25}$$

where we consider multiple road network features (e.g., route distances $d_{\mathcal{G}}(l,l')$, speed limits $v_{\mathcal{G}}(l,l')$, and road network density $den_{\mathcal{G}}(l,l')$) in $\mathcal{D}_{\mathcal{G}}(l,l')$, as shown in (5.7). *Let*

$$\mathcal{D}_{\mathcal{G}}(l,l') = \sum_{k=1}^{K} \alpha_k g_{\mathcal{G}}^k(l,l'), \tag{5.26}$$

where K is the number of considered road network features, $g_{\mathcal{G}}^k(l,l')$ is the k-th road network feature difference between l and l', and α_k is the k-th weighting factor. As the features hold the triangle inequality, we have

$$\mathcal{D}_{\mathrm{G}}(l_0,l')-\mathcal{D}_{\mathrm{G}}(l_1,l') = \sum_{k=1}^{K} \alpha_k \left(g_{\mathcal{G}}^k(l_0,l') - g_{\mathcal{G}}^k(l_1,l')\right) \le \sum_{k=1}^{K} \alpha_k g_{\mathcal{G}}^k(l_0,l_1) = \mathcal{D}_{\mathrm{G}}(l_0,l_1). \tag{5.27}$$

By following the analysis of (5.17) - (5.22) *in the proof of Theorem 5.1, the following holds*

$$\begin{cases} e^{\frac{\epsilon}{2}\mathcal{D}_{\mathcal{G}}(l_0,l')-\mathcal{D}_{\mathcal{G}}(l_1,l')} \le e^{\frac{\epsilon}{2}\mathcal{D}_{\mathcal{G}}(l_0,l_1)}; & (5.28) \\ \dfrac{\sum_{l_j\in\mathbb{R}^2} e^{\frac{\epsilon}{2}\mathcal{D}_{\mathcal{G}}(l_j,l')}}{\sum_{l_i\in\mathbb{R}^2} e^{\frac{\epsilon}{2}\mathcal{D}_{\mathcal{G}}(l_i,l')}} \le e^{\frac{\epsilon}{2}\mathcal{D}_{\mathcal{G}}(l_0,l_1)}. & (5.29) \end{cases}$$

Finally, by substituting (5.28) *and* (5.29) *into* (5.25), *we have*

$$\frac{\Pr[l'|l_0]}{\Pr[l'|l_1]} \le e^{\epsilon\mathcal{D}_{\mathcal{G}}(l_0,l_1)}, \tag{5.30}$$

which satisfies (5.24). *In other words, CR_0 and CR_1 are RN-indistinguishable*

Corollary 5.1 *The introduced CRO mechanism using route distance $d_{\mathcal{G}}(l,l')$, speed limit difference $v_{\mathcal{G}}(l,l')$, and road density difference $den_{\mathcal{G}}(l,l')$ for $\mathcal{D}_{\mathcal{G}}(l,l')$ satisfies $(\epsilon,\mathbb{R}^2)$-RN-indistinguishability.*

Proof 5.3 *The triangle inequality of route distance $d_{\mathcal{G}}(l,l')$ in road network has been proved by following the analysis of* (5.17) - (5.22) *in the proof of Theorem 5.1.*

The speed limit difference $v_{\mathcal{G}}(l,l')$ and road network density difference between two cloaking regions CR (with location l) and CR' (with location l') is as follows

$$\begin{cases} v_{\mathcal{G}}(l,l') = |v(l) - v(l')|; & (5.31) \\ den_{\mathcal{G}}(l,l') = |den(l) - den(l')|, & (5.32) \end{cases}$$

where $v(l)$ and $den(l)$ are the speed limit and road network density in cloaking region CR, respectively.

For any positive number a, b, and c, the following holds

$$|a-b| - |b-c| \le |a-c|. \tag{5.33}$$

Using the definition of $v_{\mathcal{G}}(l,l')$ in (5.31), *the above equation can be rewritten as*

$$\begin{aligned} |v(l_0) - v(l_2)| - |v(l_1) - v(l_2)| &\le |v(l_0) - v(l_1)|; \\ v_{\mathcal{G}}(l_0,l_2) - v_{\mathcal{G}}(l_1,l_2) &\le v_{\mathcal{G}}(l_0,l_1); \\ \frac{v_{\mathcal{G}}(l_0,l_2)}{\max_{\mathbb{R}^2} v_{\mathcal{G}}} - \frac{v_{\mathcal{G}}(l_1,l_2)}{\max_{\mathbb{R}^2} v_{\mathcal{G}}} &\le \frac{v_{\mathcal{G}}(l_0,l_1)}{\max_{\mathbb{R}^2} v_{\mathcal{G}}}, \end{aligned} \tag{5.34}$$

where $\max_{\mathbb{R}^2} v_{\mathcal{G}}$ *is the maximum speed limit difference in* $\mathbb{R}^2$.

Similarly, $den_{\mathcal{G}}(l, l')$ *has*

$$\frac{den_{\mathcal{G}}(l_0, l_2)}{\max_{\mathbb{R}^2} den_{\mathcal{G}}} - \frac{den_{\mathcal{G}}(l_1, l_2)}{\max_{\mathbb{R}^2} den_{\mathcal{G}}} \leq \frac{den_{\mathcal{G}}(l_0, l_1)}{\max_{\mathbb{R}^2} den_{\mathcal{G}}}. \tag{5.35}$$

We set values of α_1, α_2, *and* α_3 *larger than 0, and set other factors equal to 0. Thus,*

$$\begin{aligned}
\mathcal{D}_{\mathcal{G}}(l_0, l_2) - \mathcal{D}_{\mathcal{G}}(l_1, l_2) = \alpha_1 & \left(\frac{d_{\mathcal{G}}(l_0, l_2)}{\max_{\mathbb{R}^2} d_{\mathcal{G}}} - \frac{d_{\mathcal{G}}(l_1, l_2)}{\max_{\mathbb{R}^2} d_{\mathcal{G}}}\right) \\
& + \alpha_2 \left(\frac{v_{\mathcal{G}}(l_0, l_2)}{\max_{\mathbb{R}^2} v_{\mathcal{G}}} - \frac{v_{\mathcal{G}}(l_1, l_2)}{\max_{\mathbb{R}^2} v_{\mathcal{G}}}\right) \\
& + \alpha_3 \left(\frac{den_{\mathcal{G}}(l_0, l_2)}{\max_{\mathbb{R}^2} den_{\mathcal{G}}} - \frac{den_{\mathcal{G}}(l_1, l_2)}{\max_{\mathbb{R}^2} den_{\mathcal{G}}}\right) \\
& \leq \alpha_1 \frac{d_{\mathcal{G}}(l_0, l_1)}{\max_{\mathbb{R}^2} d_{\mathcal{G}}} + \alpha_2 \frac{v_{\mathcal{G}}(l_0, l_1)}{\max_{\mathbb{R}^2} v_{\mathcal{G}}} + \alpha_3 \frac{den_{\mathcal{G}}(l_0, l_1)}{\max_{\mathbb{R}^2} den_{\mathcal{G}}} \\
& \leq \mathcal{D}_{\mathcal{G}}(l_0, l_1).
\end{aligned} \tag{5.36}$$

$\mathcal{D}_{\mathcal{G}}(l, l')$ *with route distance* $d_{\mathcal{G}}(l, l')$, *speed limit difference* $v_{\mathcal{G}}(l, l')$, *and road density difference* $den_{\mathcal{G}}(l, l')$ *satisfies the triangle inequality in road network.*

According to the Theorem 5.2, the introduced CRO mechanism using $d_{\mathcal{G}}(l, l')$, $v_{\mathcal{G}}(l, l')$, *and* $den_{\mathcal{G}}(l, l')$ *satisfies* $(\epsilon, \mathbb{R}^2)$*-RN-indistinguishability.*

5.5 Experimental Results

In this section, we evaluate the CRO mechanism with real-world road networks[1] and the T-Drive trajectory (10,357 divers in Beijing, China) [462]. We extract two real-world road networks with different densities from the Open Street Map[2], as shown in Figs. 5.4 and 5.5. We set $\alpha_1 = 1$ in (5.8), which indicates that we use route distance $d_{\mathcal{G}}(l_i, l_j)$ as the only metric of RN-indistinguishability.

5.5.1 Location Privacy Protection in High-Density Road Network

We first compare the average AEE of the CRO mechanism with the 2D Laplace mechanisms [166, 12] in a high-density road network, as shown in Figure 5.4. The 2D Laplace mechanism treats a road network as a 2D plane and employs the Euclidean distance metrics. Different privacy-preserving requirements are characterized by privacy budgets of the cloaking regions, which are set to 0.1, 0.5, 1, 5, and 10 in the simulation. The obfuscation radii are set to 500 m and 600 m. To illustrate the advantages of considering the features

[1] There are 69239 connections and 80943 roads within the high-density road network (116.246898°≤ longitude ≤ 116.5089155°and 39.78669°≤ latitude ≤ 40.0447621°) with 83 connections and 96 roads per km^2.
There are 29989 connections and 33950 roads within the low-density road network (116.432589°≤ longitude ≤ 116.7437658°and 39.7189659° ≤ latitude ≤ 40.025068°) with 8 connections and 9 roads per km^2.

[2] Open Street Map is an open-source database of the world's geographic map. https://www.openstreetmap.org/

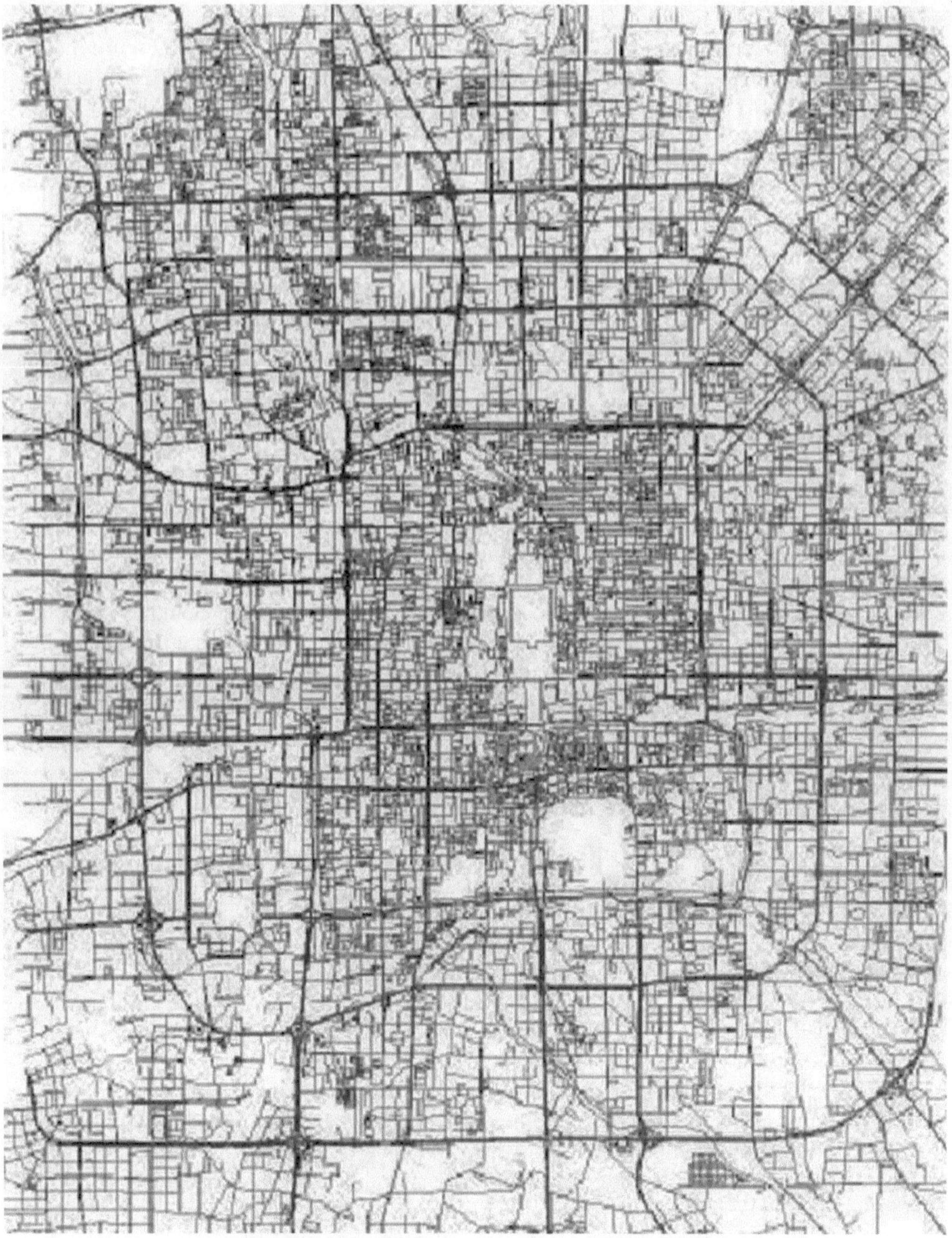

FIGURE 5.4: The road network with high density.

of road networks fairly, we compare our mechanism and the other two Laplace mechanisms based on Euclidean distance and route distance. The size of considered cloaking regions is 50 m × 50 m in the high-density road network. A CR is empty if there is no connection or road in it. The empty CR is removed. Then, we obfuscate the cloaking regions with the introduced CRO mechanism and 2D Laplace mechanisms developed in [166, 12].

Experimental results show that the CRO mechanism protects location privacy better than the 2D Laplace mechanisms (i.e., [166, 12]) in road networks, as shown in Figure 5.6. The obfuscation radii are 500 m and 600 m, respectively. The average AEE of the introduced CRO mechanism is less than that of the 2D Laplace mechanism. The average AEE of the mechanisms decreases as ϵ increases. With the introduced mechanism, a driver selects an obfuscated cloaking region, which is close to the actual location, with a higher probability of using the route distance as a metric than that of using Euclidean distance, due to (5.10). The obfuscation mechanism also has a high probability of selecting a close cloaking region as an obfuscated cloaking region in the high ϵ environment. Thus, the decreasing trend of the introduced CRO mechanism is faster than that of its counterparts. The average AEE of the CRO mechanism under the various ϵ values can be greater than the 2D Laplace mechanisms. The average AEE with Euclidean distance is much shorter than that with route distance. There are two reasons: Firstly, the adjacent cloaking regions in 2D maps may not be adjacent in road networks. Secondly, the empty cloaking regions in obfuscation areas are removed, so each obfuscation area covers an uneven number of cloaking regions.

The comparison of the average SD between the introduced CRO mechanism and the 2D Laplace mechanism is shown in Figure 5.7. The radii of the obfuscation areas are 500 m and 600 m, respectively. The average SD is measured by the shortest route distance between the actual and obfuscated cloaking regions. The CRO mechanism outperforms the 2D mechanism in terms of the average SD. This is due to the fact that the CRO mechanism has a high probability of selecting a close cloaking region as the obfuscated cloaking region. The average SD of the CRO mechanism fluctuates in Figure 5.7. The reason is that the CRO mechanism is measured by the route distance in the obfuscated cloaking regions. In the same obfuscation area, two cloaking regions with a short Euclidean distance may have a considerably long route distance.

5.5.2 Location Privacy Protection in Low-Density Road Network

We compare the CRO mechanism with the 2D Laplace mechanism [166, 12] in the low-density road network in Figure 5.5. The size of the cloaking region is set to 50 m $\times$ 50 m. We use the 2D Laplace mechanism [166, 12] to obfuscate the cloaking regions. The privacy budgets of the cloaking regions are set to 0.1, 0.5, 1, 5, and 10 for simulating different privacy-preserving requirements. The obfuscation radii are set to 500 m and 600 m.

We compare the average SD of the CRO mechanism and the 2D Laplace mechanism in low-density road networks, as shown in Figure 5.8. The CRO mechanism can achieve a shorter average SD than the 2D Laplace mechanisms in [166, 12]. The gap of the average SD with the Euclidean distance between the two curves decreases as the privacy budget increases, while that with the route distance increases with the privacy budget. The reason is that the obfuscation mechanisms have a high probability of selecting adjacent cloaking regions as obfuscated results. Thus, the CRO mechanism and 2D mechanisms have a similar Euclidean-distance-based average SD under a high-privacy budget. The 2D mechanisms use the Euclidean distance as the metric to obfuscate cloaking regions, while the introduced CRO mechanism employs the route distance. Thus, the gap in the route-distance-based average SD between the CRO mechanism and 2D mechanisms increases with the privacy budget.

The CRO mechanism achieves a longer average AEE than the 2D mechanism in low-density road networks, as shown in Figure 5.9. The CRO mechanism and 2D mechanisms have a similar average AEE when the privacy budget is high. The average AEEs of the CRO and 2D mechanisms are shorter than the average SDs of those because we consider the worst-case scenario in the experiments, as shown in (5.5).

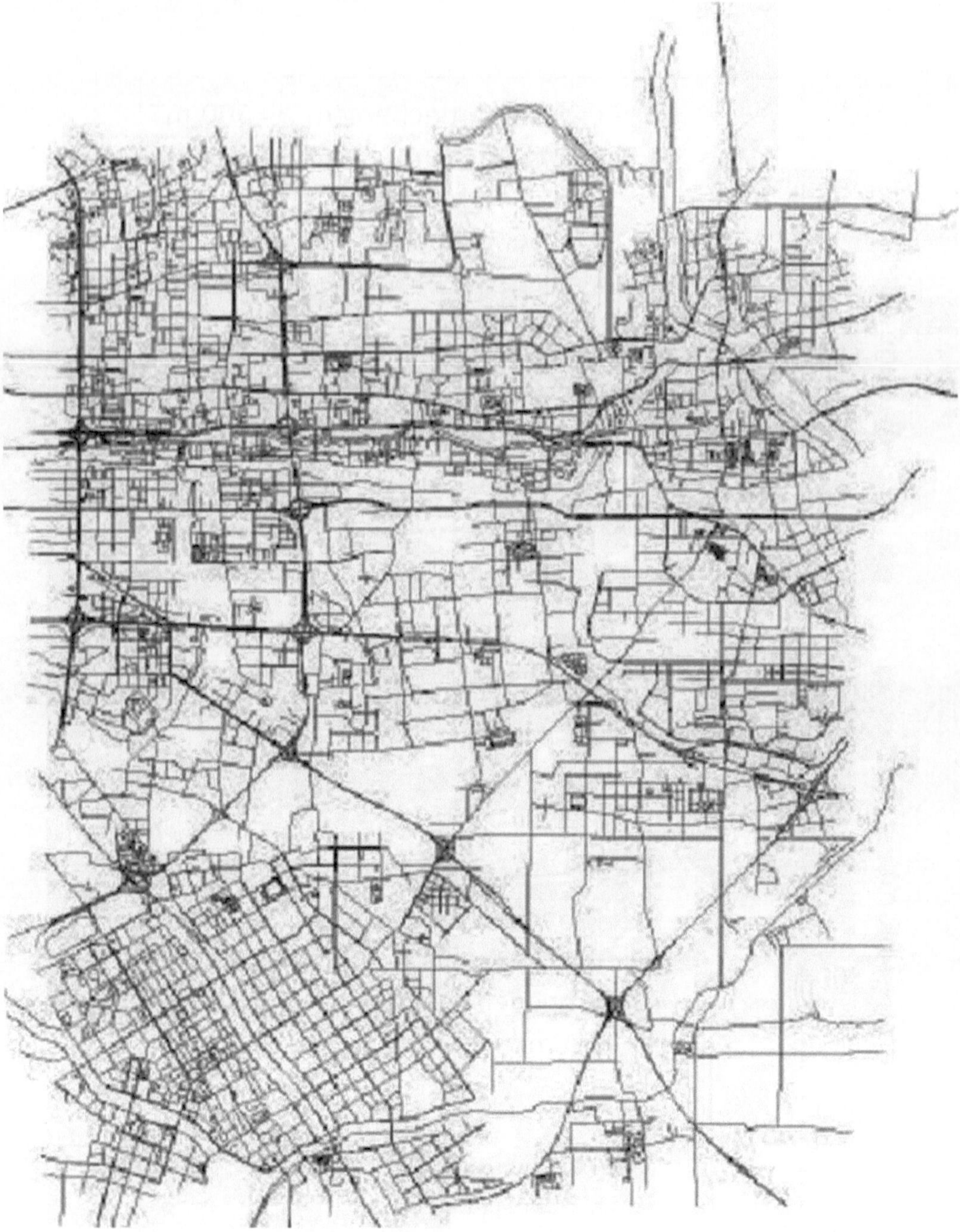

FIGURE 5.5: The road network with low density.

By comparing the results in Figures 5.6 and 5.7 with the ones in Figures 5.8 and 5.9, the introduced CRO mechanism outperforms the 2D Laplace mechanism in the high-density and low-density road networks. The introduced CRO mechanism can better protect location privacy in high-density road networks than it does in low-density road networks. The reason is that the road network density in the high-density road networks is 10 times that of the low-density road networks in our experiments. The CRO mechanism can accurately analyze the

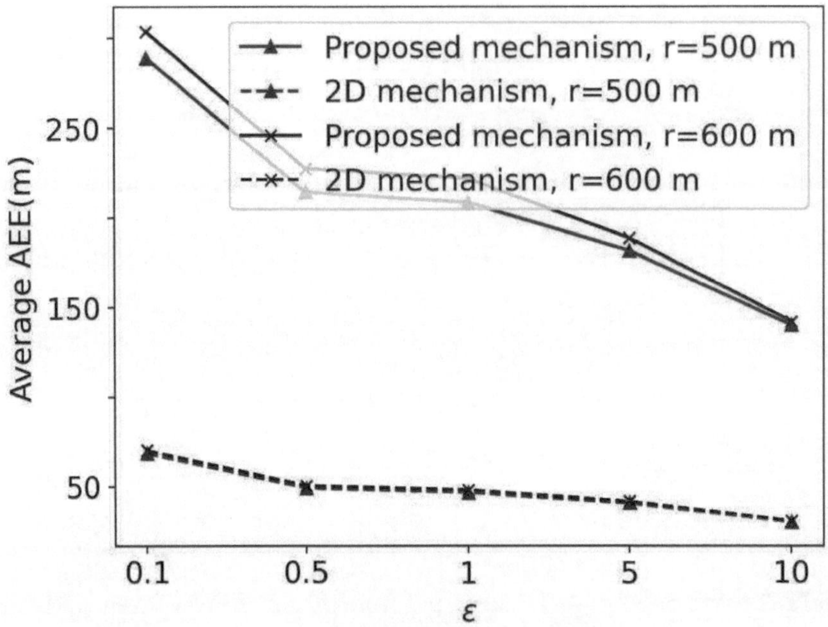

(a) Average AEE with Euclidean distance.

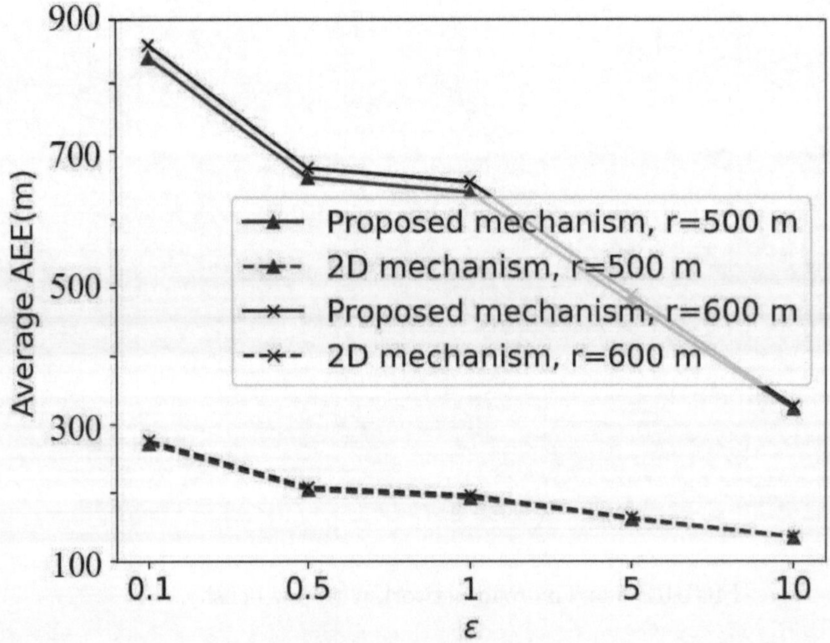

(b) Average AEE with route distance.

FIGURE 5.6: Average AEE comparisons of the introduced CRO mechanism and the 2D Laplace mechanisms in the high-density road network.

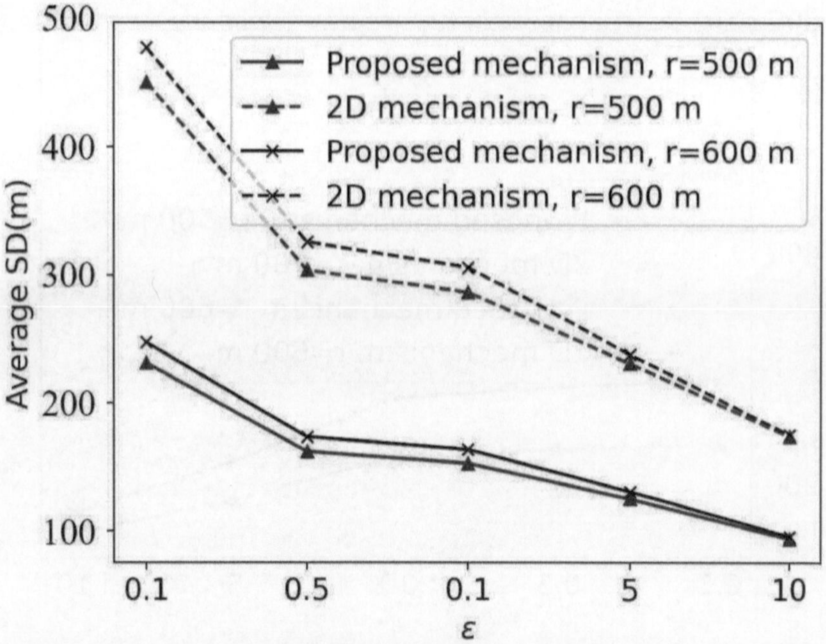

(a) Average SD with Euclidean distance.

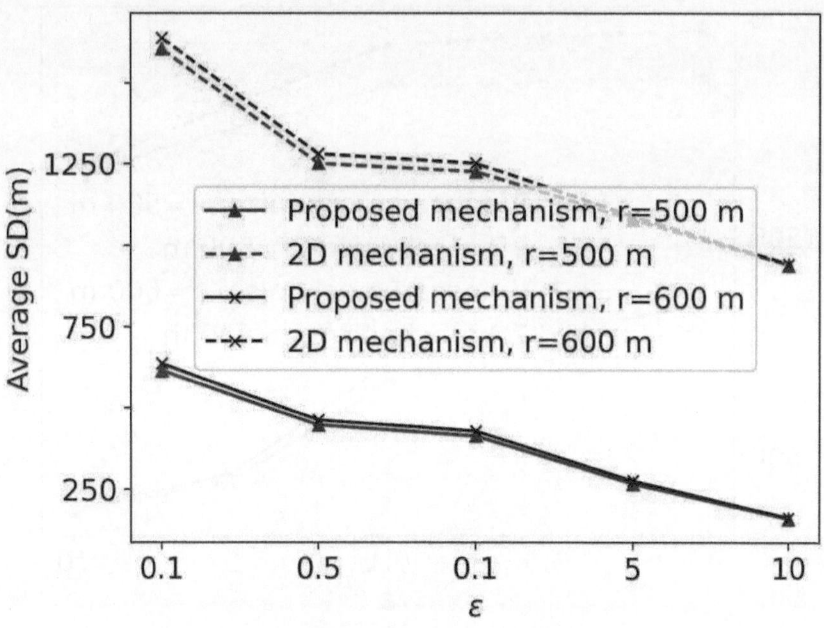

(b) Average SD with route distance

FIGURE 5.7: Average SD comparisons of the introduced CRO mechanism and the 2D Laplace mechanisms in the high-density road network.

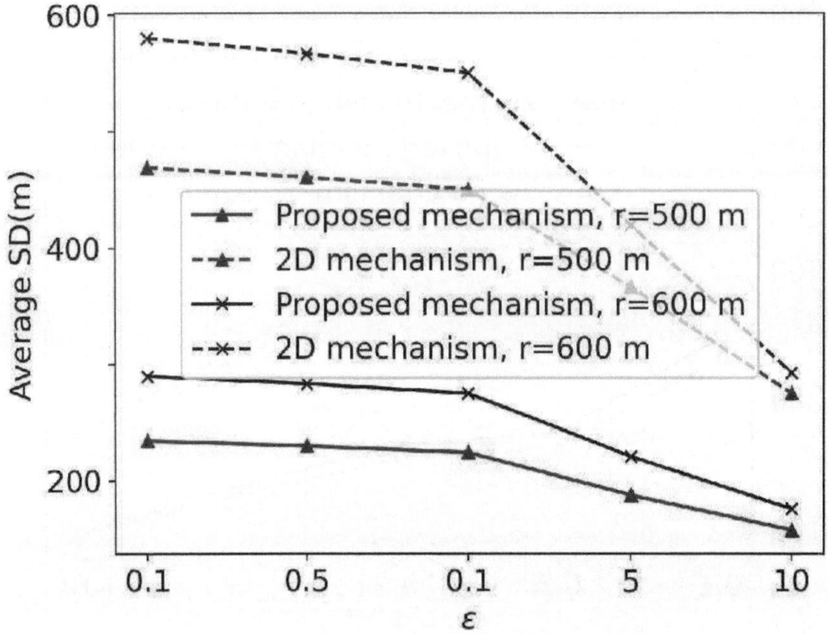

(a) Average SD with Euclidean distance

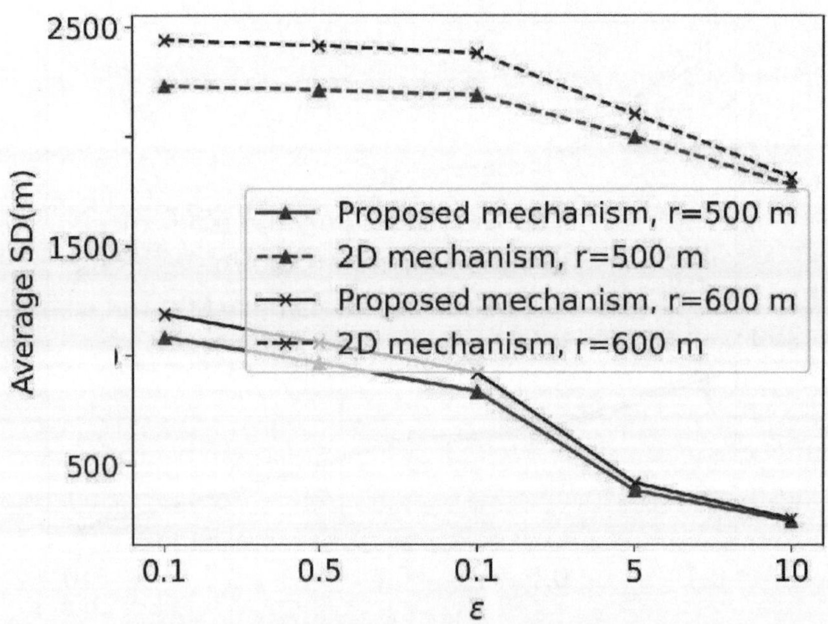

(b) Average SD with route distance

FIGURE 5.8: Average SD comparisons of the introduced CRO mechanism and the 2D Laplace mechanisms in the low-density road network.

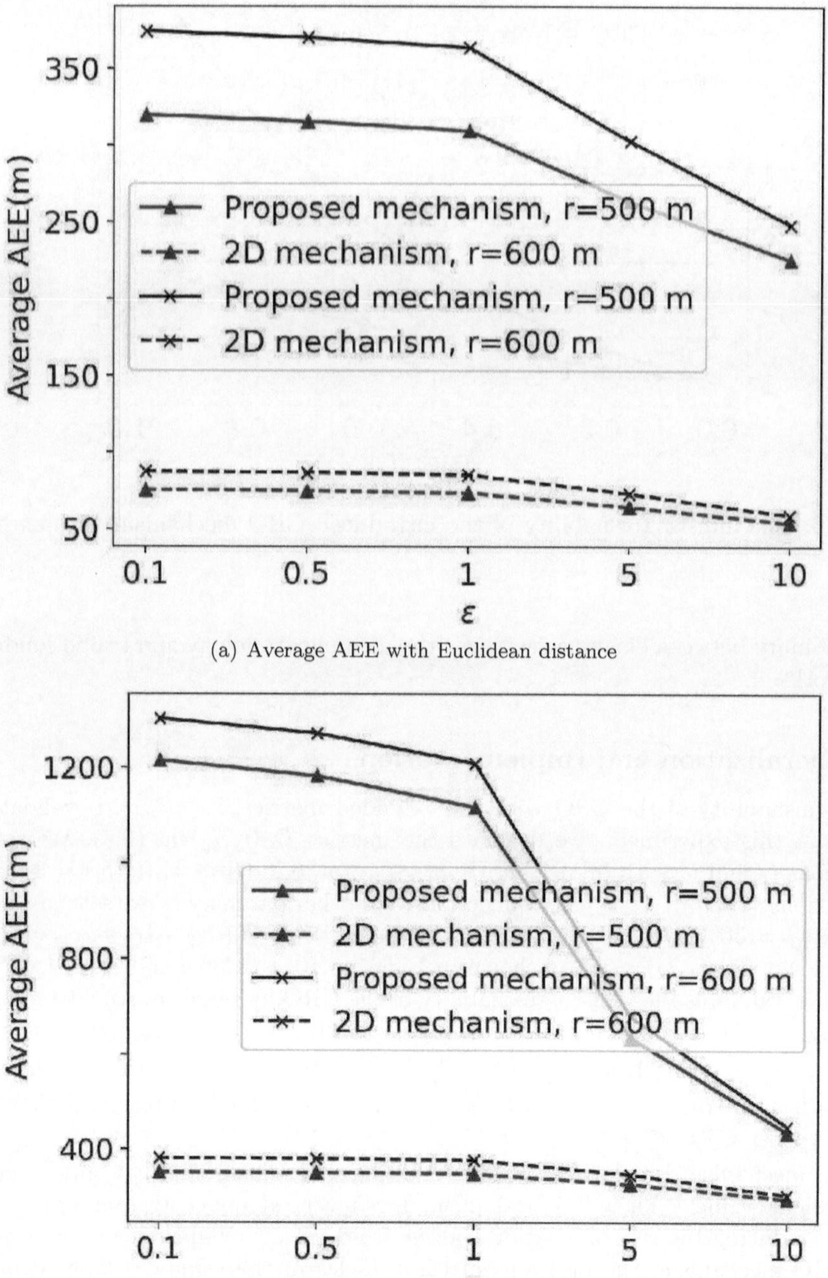

(a) Average AEE with Euclidean distance

(b) Average AEE with route distance

FIGURE 5.9: Average AEE comparisons of the introduced CRO mechanism and the 2D Laplace mechanisms in the low-density road network.

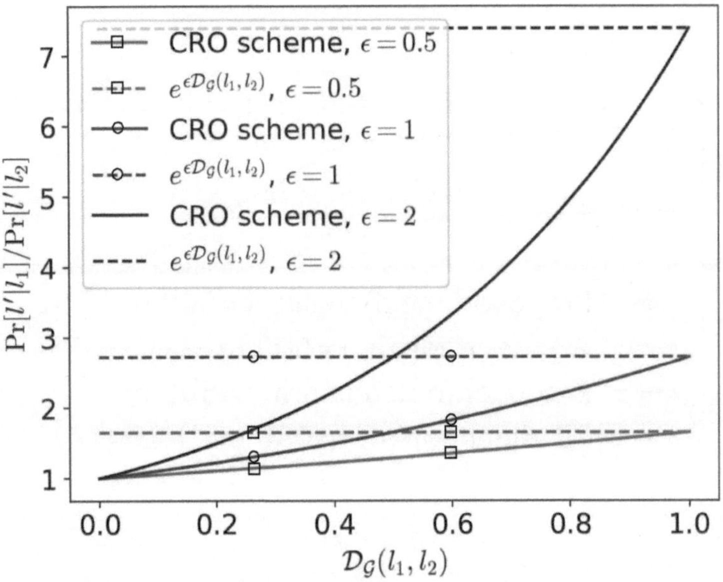

FIGURE 5.10: The indistinguishability of the introduced CRO mechanism with extended metrics.

indistinguishability between cloaking regions with a complex topology and traffic conditions in road networks.

5.5.3 Generalization and Implementation

The indistinguishability of the CRO with the extended metrics, i.e., (5.7), is validated in Figure 5.10. In this experiment, we evaluate the metrics $\mathcal{D}_{\mathcal{G}}(l_1, l_2)$ in (5.7). We set two cloaking regions l_1 and l_2 with 0 km $\leq d_{\mathcal{G}}(l_1, l_2) \leq$ 1 km, 0 km/h $\leq v_{\mathcal{G}}(l_1, l_2) \leq$ 30 km/h, and 0 connections $\leq deng_{\mathcal{G}}(l_1, l_2) \leq$ 10 connections. For other parameters, we set $\max_{\mathbb{R}^2} d_{\mathcal{G}} =$ 1 km, $\max_{\mathbb{R}^2} v_{\mathcal{G}} =$ 30 km/h, $\max_{\mathbb{R}^2} deng_{\mathcal{G}} =$ 10 connections, and $\alpha_1 + \alpha_2 + \alpha_3 = 1$. And thus, we have $0 \leq \mathcal{D}_{\mathcal{G}}(l_1, l_2) \leq 1$. As shown in Figure 5.10, the left-hand side of (5.25) is bounded, which validates the indistinguishability of the CRO mechanism. An adversary can hardly distinguish the cloaking regions that have a similar $\mathcal{D}_{\mathcal{G}}(l_1, l_2)$ from the obfuscated locations, i.e., $\frac{\Pr[l'|l_1]}{\Pr[l'|l_2]} = 1$ when $\mathcal{D}_{\mathcal{G}}(l_0, l') - \mathcal{D}_{\mathcal{G}}(l_1, l') = 0$. We also see that the left-hand side of (5.25) is approximately linear with a small ϵ, e.g., $\epsilon = 0.5$. The reason is that when $0 < x << 1$, $e^x = 1 + x + \frac{x^2}{2!} + \frac{x^3}{3!} + \cdots + \frac{x^n}{n!} \approx 1 + x$.

The CRO mechanism is a local privacy-preserving mechanism that can independently run at every single vehicle. The CRO mechanism does not need any central server to coordinate, thereby reducing the risk of location leakage from servers or during communications. With the CRO mechanism, the locations of the vehicles in the same cloaking region are uniformly processed. The location privacy-preserving capability of the introduced CRO mechanism increases with the number of vehicles in the same cloaking region. In the case of no or few neighbors, the CRO mechanism can still guarantee the drivers' location privacy by generating obfuscated locations because the drivers can locally run the CRO mechanism to protect their location data. The setting of the cloaking region can be optimized using the mechanisms developed in [251] and [448]. Vehicles in the same cloaking region can cooperate by swapping pseudonyms (e.g., [224]) or generating pseudonyms (e.g., [146]) to achieve high-privacy protection.

5.6 Conclusion

In this chapter, we introduce the RN-Indistinguishability to evaluate obfuscation-based location privacy-preserving mechanisms in road networks. We introduce the new CRO mechanism to protect the location privacy of vehicles in road networks by uniformly obfuscating the locations of vehicles in the same cloaking region. The introduced CRO mechanism has been proven to achieve RN-indistinguishability and has been validated with comprehensive experiments. We also prove that the CRO can be generalized with road network features without breaching differential privacy.

6

Cooperative Trajectory Privacy Protection

The development of vehicular networks leads to the rapid growth of location data, which potentially increases the risk of a data privacy breach. Existing studies have separately utilized pseudonyms or trajectory obfuscation to avoid privacy leakage.

This chapter proves, for the first time, that the anonymity level of pseudonym update can be transformed into a Differential Privacy (DP) expression. Hence, separately combining pseudonym update with DP can breach the data privacy protection of DP by evoking the additivity composition theorem. To address this issue, we define Trajectory-Indistinguishability (T-I) to measure the indistinguishability of vehicles in road networks. It captures trajectory obfuscation and pseudonym swapping in a DP framework. Based on the T-I, we design a new Joint Trajectory Obfuscation and Pseudonym Swapping (JTOPS) mechanism to protect the identity and trajectory privacy of vehicles on roads. With the JTOPS, vehicles can locally obfuscate their spatio-temporal driving statuses and swap their pseudonyms according to DP. The JTOPS is proven to achieve the T-I and avoid extra privacy costs, i.e., the JTOPS periodically resets the additive composition theorem of the DP process by using anonymity. It can protect the trajectory privacy of vehicles even when adversaries successfully attack the coordinators, which has not been addressed in the literature. Comprehensive experiments show that the JTOPS outperforms the existing approaches in the trajectory privacy-preserving capability and data utility.

6.1 Introduction

Vehicular applications provide various services with locations that are uploaded from vehicles [175]. The application servers can obtain and store the geolocation and driving statuses of vehicles, raising the concern of the vehicle's location privacy [191]. Adversaries can attack the application servers and eavesdrop on communications between the servers and vehicles for shared geolocation data, with which the adversaries can infer the drivers' private information, such as whereabouts, religion, job, and home address [296, 453, 485]. The adversary can also estimate the vehicles' locations with reported driving statuses, such as driving speed and direction [371]. Compared with other mobile users, the vehicles are easier to track because their trajectories are predictable in road networks [285, 255].

Obfuscation-based mechanisms have been developed to protect the location privacy of vehicles against eavesdroppers and malicious servers [219, 388, 31]. The vehicles can locally perturb their actual driving statuses by exploiting Differential Privacy (DP) mechanisms that add controllable noises to the data [452] or probabilistically selecting obfuscated candidates [224]. Then, the vehicles upload the obfuscated data to the servers for services [321],

DOI: 10.1201/9781003518044-6

such as location-based recommendations [271], entertainment, crowd sensing [129], and data analysis [441]. The state-of-the-art obfuscation-based mechanisms, such as [50] and [388], have considered spatial perturbation on two-dimensional (2D) plane maps and road networks. However, the temporal information has not been well considered in road networks by the existing obfuscation mechanisms [106]. If an adversary can track the identities (ID) of the vehicles, it can reduce the perturbation noise by linking the data of a specific vehicle ID over a long period, such as a long-observation attack [289]. Hence, the existing obfuscation-based mechanisms can hardly protect location privacy on a long-term basis [50].

Pseudonym-based mechanisms allow vehicles to utilize pseudonyms, rather than actual identities, to upload messages [36], hence protecting location privacy at the time domain. A vehicle can use multiple pseudonyms to share data in different periods [460]. The pseudonym-based mechanism can be classified into pseudonym generation and pseudonym swap. The pseudonym generation requires vehicles to keep silent in a specific region and use new pseudonyms simultaneously, while the pseudonym swap asks multiple vehicles to exchange their pseudonyms. The pseudonym-based mechanisms can provide high data utility with actual data, but the limitation is that the adversary can link multiple pseudonyms to a vehicle by using the spatial-temporal relevance among trajectories based on its prior knowledge [242, 224]. With the pseudonym swap, the vehicles exchange their pseudonyms with those that have similar trajectories, reducing the probability of accurate linkage through differential cryptanalysis compared to pseudonym generation [13]. Existing studies have shown that the more similar two vehicles' driving statuses are, the better privacy protection can be achieved [104, 418, 341].

By using existing hybrid mechanisms, vehicles report obfuscated locations generated based on their actual locations with pseudonyms [52, 389]. Pseudonym swapping relies on the spatial and temporal correlation among multiple vehicles, allowing only nearby vehicles to swap pseudonyms for location privacy and data utility. On the other hand, local obfuscation only uses the local data of a single vehicle and overlooks the spatial and temporal correlation among vehicles. It has been generally believed that pseudonym and obfuscation are two separate technologies [10, 84]; i.e., trajectory obfuscation and pseudonym swapping are processed separately and in parallel [52, 389]. Existing hybrid mechanisms separately employ pseudonym swapping and trajectory obfuscation, and typically overlook the prior knowledge in their pseudonym-swapping processes. The pseudonym-swapping process of the existing hybrid mechanisms could undergo a low data utility and privacy-preserving capability. For example, the pseudonym-swapping process without using differential privacy can be attacked by the adversary with prior knowledge [224]. As a matter of fact, we reveal analytically in this chapter that pseudonym swaps can be considered a differential privacy process for any two identities sharing the same pseudonym-swapping candidate set.

In this chapter, we introduce a novel Joint Trajectory Obfuscation and Pseudonym Swapping (JTOPS) mechanism to prevent adversaries from learning actual driving statuses and matching their prior knowledge of vehicles. We start by designing a new distance metric to measure the difference among spatio-temporal driving statuses, and defining Trajectory-Indistinguishability (T-I) to evaluate the privacy-preserving ability against adversaries who have the prior knowledge of the drivers' historical trajectories. DP is leveraged to jointly obfuscate the driving statuses and swap the drivers' pseudonyms. The vehicles firstly obfuscate their driving statuses, according to the distance between their actual statuses and candidates. Then, they swap their pseudonyms, according to the swapping probabilities calculated based on the distances among their obfuscated statuses. The JTOPS is proven to achieve T-I under the Global Passive Adversaries (GPAs) model with full prior knowledge of the vehicles at the adversary. Additionally, the introduced JTOPS combines two differential privacy processes, without introducing the additivity composition theorem of multiple ε-DP processes [137]. The introduced mechanism can still effectively disguise the vehicles' driving statuses when their swapped pseudonyms are exposed.

The topics of this chapter are summarized as follows:

(1) This chapter demonstrates that pseudonym swaps can be considered a differential privacy process for any two identities sharing the same pseudonym-swapping candidate set. We analytically prove that jointly using pseudonym swapping and obfuscation in vehicular networks can achieve higher privacy protection than separately using them.

(2) We introduce a new unified privacy-preserving measure, i.e., T-I, treating trajectory obfuscation and pseudonym-based mechanisms as a holistic process. T-I extends the applicability of ε-differential privacy (DP) by quantifying the distinguishability between vehicles based on their historical information in the time domain and current information in the spatial domain.

(3) Building upon T-I, we introduce a novel Joint Trajectory Obfuscation and Pseudonym Swapping (JTOPS) mechanism joining pseudonym swapping and trajectory obfuscation with a crafted criterion. The introduced mechanism is proven to combine two differential privacy processes, without introducing the additivity composition theorem of multiple ε-DP processes.

(4) The introduced pseudonym swapping does not require vehicles to disclose their private data, as the required parameters are computed locally and then transmitted to the coordinator. Thus, the knowledge of the coordinator, which assists in pseudonym swapping, is limited, ensuring the introduced pseudonym-swapping process can effectively resist collusion attacks.

We experimentally assess the JTOPS on the real-world road network and trajectory datasets, compared to three state-of-the-art vehicle privacy-preserving mechanisms, i.e., those developed in [402], [255], and [224]. The superiority of the JTOPS is demonstrated in terms of privacy preservation and data utility in the cases of GPAs and collusion attacks.

The rest of this chapter is organized as follows. Section 6.2 reviews the related works. Section 6.3 presents the system model. Section 6.4 analyzes the introduced T-I and JTOPS mechanisms. The experimental evaluation of the mechanism is provided in Section 6.5, followed by conclusions in Section 6.6. The notation used in the chapter is collated in Table 6.1.

6.2 Related Work

Obfuscation and anonymity have been primarily adopted to protect location privacy [196]. Obfuscation mechanisms add DP noises to actual location data or probabilistically select obfuscation candidates to reduce the precision of location data [415], while anonymity mechanisms allow vehicles to use pseudonyms to hide their real identities in different periods [191].

Most existing obfuscation-based mechanisms protect actual information by utilizing DP, making private information indistinguishable from adversaries [45]. Andrés *et al.* [31] developed Geo-Indistinguishability (Geo-I) based on DP, which employed the Laplace mechanism to realize Geo-I on a two-dimensional (2D) plane. The Geo-I reports obfuscated locations surrounding the actual location based on the obfuscation probability distribution, but the Geo-I can generate off-road obfuscated locations that compromise the data utility. The Graph-Exponential Mechanism (GEM) evaluates the privacy-preserving capability of the

TABLE 6.1: Summary of notation.

Notation	Description
$M, \hat{M}$	A sequence of actual/obfuscated driving statuses
$m, \hat{m}$	An actual/obfuscated driving status
$\bar{\mathcal{M}}$	The sequence of obfuscation candidate
$\bar{m}$	An obfuscation candidate
ε	Privacy budget
id	The pseudonym of vehicle
α_i	The i-th weight factor
T	The time of pseudonym swap
T-	The period of time before time T
T+	The period of time after time T
$(\hat{M}_a, id_a)$	The uploaded messages of v_a
$\mathcal{S}$	The pseudonym pool
β	Any driving status
λ	Data Utility
γ	Pseudonym Utility

Geo-I in road networks [388]. The GEM sets the connections (such as turns, intersections, bifurcates, and joins) in a road network as the obfuscation candidates. The mechanism maps the vehicles' locations to the connections and obfuscates the connections based on the Geo-I. Inspired by [31] and [388], Ma *et al.* [255] improved the Geo-I to Road Network-Indistinguishability (RN-I) by using route distance between two road network locations, with which the vehicles on road networks are indistinguishable to an adversary. These existing obfuscation-based mechanisms only use location data to measure the distance between two vehicles, yet overlook other road network spatial-temporal driving statuses. Moreover, the existing obfuscation-based mechanisms allow the vehicles to use their actual identities, leading to vulnerabilities to long-observation attacks with which an adversary can achieve a highly accurate estimation by observing communications of the target vehicle in a long period [289].

The pseudonym-based mechanisms allow vehicles to use unique identifications, rather than actual identities, in communications, and update the identifications periodically. Pseudonym swap and pseudonym generation are the two most popular mechanisms among pseudonym-based mechanisms to update pseudonyms. The pseudonym generation mechanisms create new pseudonyms with which multiple vehicles replace their old pseudonyms, while the pseudonym swapping mechanisms assist vehicles in exchanging their pseudonyms. The pseudonym-swapping mechanisms, which can provide higher data utility and less storage overhead than the pseudonym generation, have gained more attention from researchers [224]. Li *et al.* [226] pointed out that pseudonym-based mechanisms can provide a high privacy-preserving capability when a spatial-temporal context is considered. The work in [224] develops a pseudonym-swapping mechanism by using DP, with which the vehicles swap pseudonyms probabilistically based on the driving status similarity. Therefore, the adversary cannot gain more knowledge by observing the pseudonym-swapping process or referring to historical trajectories. The existing pseudonym-based mechanisms overlook the adversary with prior knowledge of the vehicles, and the privacy-preserving capability is low in such a case.

Existing hybrid mechanisms separately use obfuscation and pseudonyms. Ullah *et al.* [402] developed a multiple-level mechanism by allowing vehicles to upload multiple messages with different locations and pseudonyms. The mechanism uses the actual locations of other vehicles as dummy locations. A vehicle randomly reports multiple dummy locations and the corresponding pseudonyms with its actual location and pseudonym to the LBS, indicating that the uploaded messages of the vehicle include information about other vehicles. In the mechanism, the pseudonym and dummy location of a vehicle are selected at the same time. The work [51] separately uses pseudonym and obfuscation in a cloud-enabled internet where vehicles synchronously generate new pseudonyms and then report the actual or obfuscated locations. The authors of [51] only discuss the pseudonym generation algorithm, which is independent of the obfuscation.

In vehicular networks, cryptography-based mechanisms are commonly used due to their lightweight encryption that minimizes delays and mitigates negative impacts such as frequent disconnections, instability, and handshake failures caused by the unique features of vehicles [7, 233]. To reduce the communication consumption of Zero-Knowledge Proof (ZKP), Li et al. [221] introduced an improved interactive ZKP using blockchain network management. The authors presented a blockchain gateway that enables the switching of vehicles between adjacent blockchains without revealing sensitive data, thereby enhancing privacy protection. Khodaei et al. [194] developed a mix-zone-based mechanism that encrypts vehicle information within a zone using a set of virtual vehicles. This mechanism enhances privacy protection with limited additional computation and communication consumption. However, existing cryptography-based mechanisms may not be suitable for high-mobility real-time Location-Based Services (LBS) in vehicular networks, as low latency and delay in Vehicle-to-Everything (V2X) communication are crucial requirements [92].

To the best of our knowledge, the mechanism introduced herein is the first joint mechanism that protects location privacy with both pseudonym swapping and trajectory obfuscation in road networks. In the JTOPS, the pseudonyms of vehicles are swapped following the probabilities generated based on the obfuscated driving statuses. By jointly using trajectory obfuscation and pseudonym swapping, the JTOPS provides excellent location privacy protection under GPAs with prior knowledge of vehicles or under collusion attacks.

6.3 System Model

This section provides the considered system model, including the components of the system and the important conceptions.

6.3.1 Vehicular Network

A vehicular network consists of three types of entities: Trusted Authority (TA), coordinators, and vehicles, as shown in Figure 6.1.

6.3.1.1 Trusted Authority (TA)

We assume that the TA, which has unlimited resources, is managed by state representatives [390]. The TA generates a legal pseudonym for each vehicle and stores its vehicle-pseudonym mapping for potential future accident forensics and investigation.

6.3.1.2 Coordinators

The coordinators act as intermediate communication interfaces between the vehicles and the TA [390]. In our model, the coordinators assist in the pseudonym-swapping process

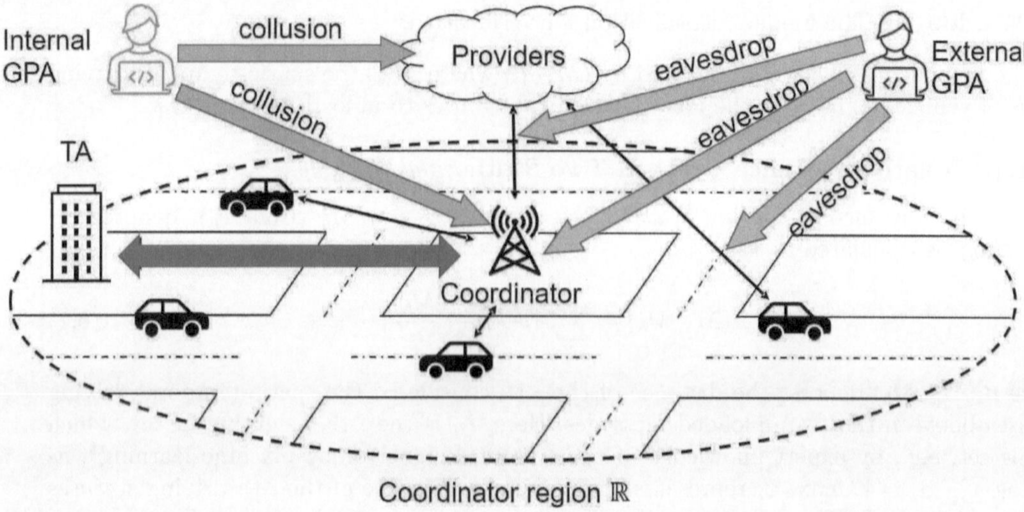

FIGURE 6.1: System and adversary model. The vehicular network consists of three parties: TA, coordinator, and vehicles.

within their network coverage regions and upload the new pseudonym-vehicle mapping to TA. The coordinators, which might collude with adversaries, are not trusted.

- **Coverage Area $\mathbb{R}$:** Each coordinator covers a region $\mathbb{R}$ in road networks, e.g., the circle in Figure 6.1, which does not overlap with other coverage regions. The vehicles in $\mathbb{R}$ swap their pseudonyms and share the same obfuscation candidates to protect their location privacy.
- **Pseudonym Pool $\mathcal{S}$:** When a coordinator assists the vehicles in swapping pseudonyms, the coordinator puts the pseudonyms of the vehicles that participate in the pseudonym swapping into a pseudonym pool $\mathcal{S}$. The number of vehicles in $\mathcal{S}$ is no larger than that of vehicles in $\mathbb{R}$. Before the pseudonym swapping, the coordinator records the recently uploaded driving statuses of the vehicles in $\mathcal{S}$ to ensure the recorded driving statuses of each vehicle have the same length.

6.3.1.3 Vehicles

The vehicles wish to enjoy location-based services but distrust the service providers. The vehicles employ pseudonym swapping and trajectory obfuscation to protect their identities and trajectory information, respectively. The vehicles share their obfuscated driving statuses (e.g., locations, speeds, and directions) with the coordinators.

A vehicle v uploads messages that contain time-series spatio-temporal driving statuses, for driving security, driving assistant, and entertainment applications [390]. The spatio-temporal information is protected by the trajectory obfuscation mechanism and uploaded with pseudonyms. The information in the uploaded messages M is as follows:

- **Spatio-Temporal Driving Statuses:** In this chapter, we consider the spatio-temporal driving statuses of the vehicles, e.g., location, speed, direction, and the driving time of v in $\mathbb{R}$. The uploaded messages contain multiple driving statuses at different times and are denoted by $M = \{m_1, \cdots, m_i, \cdots, m_k\}$, where $m_i \in M$ is the i-th driving statuses at time t and contains status like location loc, speed spe, and direction dir, i.e., $m_i = \{loc,\ spe,\ dir, \cdots\}$.

- **Pseudonym:** The unique identification for verification.

The uploaded messages are denoted by (M, id), where M is the spatio-temporal driving status of vehicle v, and id is the identification that v uses to upload its messages.

6.3.1.4 Relative Distance between Two Statuses $d(M_a, M_b)$

The relative distance of driving statuses M_a (from v_a) and M_b (from v_b), denoted by $d(M_a, M_b)$, is calculated as

$$d(M_a, M_b) = \sum_i \alpha_i \frac{\beta_i(M_a, M_b)}{\max_{\mathbb{R}} \beta_i}, \tag{6.1}$$

where $\beta_i(M_a, M_b)$ denotes the distance of the i-th considered the spatio-temporal driving statuses between the two uploaded messages. Here, α_i is the i-th weighting factor, which can be set, e.g., by expert knowledge of the system designer or by machine learning[1], as long as $\sum_{\forall i} \alpha_i = 1$. $\max_{\mathbb{R}} \beta_i$ represents the maximum distance of the i-th driving statuses of the vehicles in $\mathbb{R}$. The calculation of each considered driving status is as follows.

Corollary 6.1 *$\mathcal{D}(v_a, v_b)$ with location distance $\beta_l(M_a, M_b)$, speed distance $\beta_s(M_a, M_b)$, direction distance $\beta_r(M_a, M_b)$, and driving time distance $\beta_t(M_a, M_b)$ satisfies triangle inequality in a road network.*

Proof 6.1 *First, we prove that the vector-type driving statuses, such as route distance, satisfy the triangle inequality in road networks, i.e.,* for any three vehicles v_a, v_b, and v_c in a road network, $\beta_l(M_a, M_c) - \beta_l(M_b, M_c) \leq \beta_l(M_a, M_b)$ holds as follows.

If v_b is along the shortest route between v_a and v_c, we have

$$\beta_l(M_a, M_c) - \beta_l(M_b, M_c) = \beta_l(M_a, M_b). \tag{6.2}$$

If v_b is not along the shortest route between v_a and v_c, by assuming $\beta_l(M_a, M_c) - \beta_l(M_b, M_c) > \beta_l(M_a, M_b)$, we have

$$\beta_l(M_a, M_c) > \beta_l(M_a, M_b) + \beta_l(M_b, M_c), \tag{6.3}$$

which indicates that the route between v_a and v_c is longer than the route between v_a and v_c via v_b. Then, the shortest route between v_a and v_c should pass v_b. This contradicts with the assumption of $\beta_l(M_a, M_c) - \beta_l(M_b, M_c) > \beta_l(M_a, M_b)$. If v_b is not along the shortest route between v_a and v_c, the following inequality holds

$$\beta_l(M_a, M_c) - \beta_l(M_b, M_c) \leq \beta_l(M_a, M_b). \tag{6.4}$$

Combining (6.2) *and* (6.4), *the triangle inequality of route distance in road networks is proved, i.e.,*

$$\frac{\beta_l(M_a, M_c)}{\max_{\mathbb{R}} \beta_l} - \frac{\beta_l(M_b, M_c)}{\max_{\mathbb{R}} \beta_l} \leq \frac{\beta_l(M_a, M_b)}{\max_{\mathbb{R}} \beta_l}, \tag{6.5}$$

where $\max_{\mathbb{R}} \beta_l$ is the maximum route distance in region $\mathbb{R}$.

The numeric type of driving statuses, such as speed distance $\beta_s(M_a, M_b)$, direction distance $\beta_r(M_a, M_b)$, and the driving time distance $\beta_t(M_a, M_b)$ between any two vehicles v_a and v_b is considered as follows

$$\beta_j(M_a, M_b) = |\beta_j(M_a) - \beta_j(M_b)|, \tag{6.6}$$

[1] In the experiments, all alpha values are set to be equal.

where $\beta_j(M_a)$ is the j-th numeric-type driving status of vehicle v_a.

Based on the absolute value inequality and (6.6), it follows

$$|\beta_j(M_a)-\beta_j(M_b)|-|\beta_j(M_b)-\beta_j(M_c)|\leq|\beta_j(M_a)-\beta_j(M_b)|; \quad \beta_j(M_a,M_b)-\beta_j(M_b,M_c)\leq\beta_j(M_a,M_b), \tag{6.7}$$

where M_c is a sequence of driving statuses of any vehicle v_c.

Thus, we have

$$\frac{\beta_j(M_a,M_c)}{\max_{\mathbb{R}}\beta_j}-\frac{\beta_j(M_b,M_c)}{\max_{\mathbb{R}}\beta_j}\leq\frac{\beta_j(M_a,M_b)}{\max_{\mathbb{R}}\beta_j}, \tag{6.8}$$

where $\max_{\mathbb{R}}\beta_j$ is the maximum distance of β_j in $\mathbb{R}$. Thus,

$$\mathcal{D}(v_a,v_c)-\mathcal{D}(v_b,v_c)\leq\sum_i\alpha_i\frac{\beta_i(M_a,M_b)}{\max_{\mathbb{R}}\beta_i}=\mathcal{D}(v_a,v_b). \tag{6.9}$$

In other words, $\mathcal{D}(v_a,v_b)$ satisfies triangle inequality in road networks.

For any reported driving statuses $(\hat{M}_x^{T\text{-}},\hat{M}_x^{T+})$ from any vehicle v_c, as the driving statuses hold the triangle inequality and $0\leq\mathcal{D}(v_a,v_b)\leq1$, we have

$$\begin{cases}\mathcal{D}(v_b,v_c)-\mathcal{D}(v_a,v_c)\leq\mathcal{D}(v_a,v_b);\\ \mathcal{D}(v_a,v_c)-\mathcal{D}(v_b,v_c)\leq\mathcal{D}(v_a,v_b).\end{cases} \tag{6.10}$$

In this chapter, we consider the relative distance between the spatio-temporal driving statuses, i.e., the location distance $\beta_l(M_a,M_b)$, the speed distance $\beta_s(M_a,M_b)$, the direction distance $\beta_r(M_a,M_b)$, and the driving time distance $\beta_t(M_a,M_b)$.

6.3.2 Trajectory Obfuscation and Pseudonym Swapping

We jointly utilize pseudonym swapping and trajectory obfuscation to protect the identity and location privacy of vehicles, where the vehicles swap their pseudonyms according to the trajectory obfuscation results. When the vehicles drive in $\mathbb{R}$, they locally obfuscate their actual driving statuses for location privacy. If a vehicle needs to swap its pseudonym, it uploads its current pseudonym to $\mathcal{S}$. The vehicle then calculates the distance between its actual driving statuses and those of the other vehicles. Given the distance, the coordinator assists the vehicles in swapping their pseudonyms and returns the swapped pseudonyms to the vehicles. The vehicles use the new pseudonyms to upload their obfuscated driving statuses until the next pseudonym swap.

6.3.3 Adversary Model

We adopt the adversary model from previous studies [224, 255], and [433], which includes both external and internal GPAs, as depicted in Figure 6.1. These GPAs can infer the driving status of the vehicles by analyzing the obtained messages and subsequently tracking the vehicles. External GPAs can operate independently of the vehicular networks and eavesdrop on the uploaded messages, whereas internal GPAs are part of the vehicular networks and may collude with the servers to obtain the uploaded messages. The GPAs have complete knowledge of the historical driving statuses of the vehicles, the trajectory obfuscation strategy, and the pseudonym-swapping strategy.

To accurately track a vehicle, an adversary can infer the actual driving status of the vehicle and then estimate the real identity of the vehicle. The adversary can obtain the obfuscated driving statuses of each pseudonym. By linking the new and old pseudonyms, the adversary can have a set of obfuscated driving statuses of a target, i.e., $\hat{M}$. With $\hat{M}$, the adversaries can infer the probability that $\hat{M}$ is obfuscated by the actual driving statuses M in $\mathbb{R}$. The adversary then can match the target vehicle with its prior knowledge and the inferred probability.

By considering the adversary model and the message processing, we use the following three metrics to evaluate the introduced mechanism.

6.3.3.1 Adversary's Success Rate

The Adversary's Success Rate (ASR), i.e., the probability of successfully linking the uploaded messages $\left((\hat{M}_x^{T+}, id_x^{T+}), (\hat{M}_x^{T-}, id_x^{T-})\right)$ to v_a with pseudonym id_a, is given by

$$\begin{aligned}&\Pr[id_a|((\hat{M}_x^{T+}, id_x^{T+}), (\hat{M}_x^{T-}, id_x^{T-}))]\\ =&\frac{\Pr[((\hat{M}_x^{T+}, id_x^{T+}), (\hat{M}_x^{T-}, id_x^{T-}))|id_a]}{\sum\limits_{id_c \in \mathcal{S}} \Pr[((\hat{M}_x^{T+}, id_x^{T+}), (\hat{M}_x^{T-}, id_x^{T-}))|id_c]},\end{aligned} \tag{6.11}$$

where $(\hat{M}_x^{T-}, id_x^{T-})$ and $(\hat{M}_x^{T+}, id_x^{T+})$ are any messages in $\mathbb{R}$ before and after swapping the pseudonyms at time T. id_c is the pseudonym of any vehicle $v_c \in \mathcal{S}$.

6.3.3.2 Data Utility

We utilize the driving statuses to measure the data utility of the obfuscated driving statuses. Given a set of the obfuscated driving statuses $\hat{M}_a = \{\hat{m}_1, \cdots, \hat{m}_k\}$ from a vehicle v_a with the actual driving statuses $M_a = \{m_1, \cdots, m_k\}$, the Average Data Utility (ADU) of vehicle v_a is

$$\lambda(v_a) = \frac{1}{k}\left(1 - d(M_a, \hat{M}_a)\right), \tag{6.12}$$

where $0 \leq \lambda(v_a) \leq 1$. A high $\lambda(v_a)$ indicates that the obfuscated statuses are closer to the actual statuses; in other words, $\hat{M}_a$ provides a high data utility and quality of services.

6.3.3.3 Pseudonym Utility

The Pseudonym Utility (PU) has been overlooked in the existing pseudonym-based mechanisms. For pseudonym generation, the PU of a pseudonym is zero because the old pseudonym is abandoned. For a pseudonym swap, the PU of a pseudonym could be measured by the distance of driving statuses after swapping between the two vehicles, which use the pseudonym before and after swapping. A high PU indicates that the data uploaded by the pseudonym is similar, which means a high privacy-preserving capability and data utility of driving statuses. The PU can be defined as follows.

Given a pseudonym id_x, v_a is the vehicle that uses id_x before the pseudonym swapping, and v_b is the vehicle that uses id_x after the pseudonym swapping. The PU of id_x is given by

$$\gamma(id_x) = 1 - d(\hat{M}_a^{T+}, \hat{M}_b^{T+}), \tag{6.13}$$

where $\hat{M}_a^{T+}$ and $\hat{M}_b^{T+}$ are the obfuscated driving statuses of v_a and v_b after swapping the pseudonyms at time T.

In (6.13), the pseudonym utility is calculated based on the driving statuses after the pseudonym swapping of the two vehicles by considering the time-sequence trajectory. For example, if a vehicle does not change its pseudonym (id_x), the pseudonym utility would be 1, as the trajectory with the pseudonym id_x is reported as expected. However, if the pseudonym id_x is swapped from vehicle v_a to vehicle v_b, the pseudonym utility should be calculated based on the similarity between the new trajectory information of v_b (after the pseudonym swapping, i.e., $T+$) and what is expected based on the new trajectory information of v_a (after the pseudonym swapping, i.e., $T+$). The coordinator can flexibly decide the period of the pseudonym swapping using various methods, e.g., machine learning, to find the optimum value of T.

6.4 Joint Trajectory Obfuscation and Pseudonym Swapping Mechanism

In this chapter, the introduced driving status obfuscation mechanism is based on differential privacy, which has a privacy threshold due to the additivity composition theorem of reusing ε-DP. The privacy threshold can be exhausted if a vehicle obfuscates its driving status multiple times. The coordinator periodically broadcasts pseudonym-swapping requests, and vehicles can decide whether to join the pseudonym swap based on their requirements. If a vehicle's privacy threshold is nearly exhausted, it will join the pseudonym-swapping process upon receiving the broadcast from the coordinator. The coordinator retains all pseudonyms until the vehicle leaves the region, while the vehicle only needs to keep its current pseudonym until it joins the next pseudonym-swapping process. Once the pseudonym of a vehicle is swapped, it keeps its new pseudonym, and the coordinator uploads the new pseudonym-vehicle mapping information of the vehicle to the Trusted Authority (TA). This process ensures that the pseudonym-vehicle mapping information is updated in a timely manner and the privacy of the vehicles is maintained throughout the pseudonym-swapping process.

6.4.1 Trajectory-Indistinguishability

The T-I considers the general spatio-temporal driving statuses of continuous trajectories in vehicular networks. The introduced T-I generalizes the metric of ε-DP in vehicular networks without increasing the upper bound of the distinguishability between two vehicles and is defined as follows.

Definition 6.1 (T-I) *A mechanism satisfies $(\varepsilon, \mathbb{R})$-T-I if and only if, for any two vehicles v_a and v_b in $\mathbb{R}$, which swap their pseudonyms with other vehicles in $\mathbb{R}$, the probabilities that the uploaded message $(\hat{M}_x^{T+}, id_x^{T+})$ generated by v_a and v_b have*

$$\frac{\dfrac{\Pr[(M_a^{T-}, id_a^{T-})|(\hat{M}_x^{T+}, id_x^{T+})]}{\sum_{v_c \in \mathbb{R}} \Pr[(M_c^{T-}, id_c^{T-})|(\hat{M}_x^{T+}, id_x^{T+})]}}{\dfrac{\Pr[(M_b^{T-}, id_b^{T-})|(\hat{M}_x^{T+}, id_x^{T+})]}{\sum_{v_c \in \mathbb{R}} \Pr[(M_c^{T-}, id_c^{T-})|(\hat{M}_x^{T+}, id_x^{T+})]}} \leq e^{\varepsilon \mathcal{D}(v_a, v_b)} \frac{\Pr[(M_a^{T-}, id_a^{T-})]}{\Pr[(M_b^{T-}, id_b^{T-})]}, \tag{6.14}$$

where ε is the privacy budget, and pseudonym swapping happens at time T. M_a^{T-} and M_b^{T-} are the actual driving statuses of v_a (with pseudonym id_a^{T-}) and v_b (with pseudonym id_b^{T-}) before T, respectively. id_x is any pseudonym in $\mathbb{R}$, and $\hat{M}_x^{T+}$ are the driving statuses uploaded with id_x. $\Pr[(M_a^{T-}, id_a^{T-})]$ is the probability that the actual driving statuses generated by v_a

have appeared in the historical information according to the prior knowledge [237]. The prior knowledge could be gained by observing the vehicles before they are protected. $(M_c^{T\text{-}}, id_c^{T\text{-}})$ *is the actual message before a pseudonym swap generated by any vehicle* $v_c \in \mathbb{R}$. $\mathcal{D}(v_a, v_b)$ *specifies the distance between* v_a *and* v_b*, as given by*

$$\mathcal{D}(v_a, v_b) = d(M_a^{T+}, M_b^{T+}) + d(M_a^{T\text{-}}, M_b^{T\text{-}}). \tag{6.15}$$

Here, $d(M_a^{T\text{-}}, M_b^{T\text{-}})$ *and* $d(M_a^{T+}, M_b^{T+})$ *are the driving status distance of the two vehicles before and after* T*, respectively.*

Based on Bayes' theorem, (6.14) can be written as

$$\frac{\Pr[(\hat{M}_x^{T+}, id_x^{T+})|(M_a^{T\text{-}}, id_a^{T\text{-}})]}{\Pr[(\hat{M}_x^{T+}, id_x^{T+})|(M_b^{T\text{-}}, id_b^{T\text{-}})]} \leq e^{\varepsilon\mathcal{D}(v_a, v_b)}. \tag{6.16}$$

In other words, for any sequence of messages $(\hat{M}_x^{T+}, id_x^{T+})$ after a pseudonym swap, v_a and v_b are indistinguishable.

6.4.2 Mechanism Details

In JTOPS, the vehicles locally obfuscate the spatio-temporal driving statuses. Periodically, the vehicles swap their pseudonyms with the assistance of the coordinator and then use the new pseudonyms to upload their obfuscated driving statuses. The JTOPS is described as follows.

6.4.2.1 Registration

When the vehicles first join vehicular networks, they register with the TA, and the TA allocates an initial pseudonym to each vehicle through secure channels and stores the pseudonym-vehicle mappings. Once a vehicle enters a new $\mathbb{R}$, it sends its pseudonym to the coordinator for notification. The coordinator records the pseudonym and keeps it until the vehicle leaves $\mathbb{R}$.

6.4.2.2 Trajectory Obfuscation

When driving in $\mathbb{R}$, the vehicles periodically upload their messages with pseudonyms and obfuscated driving statuses. There are two steps in trajectory obfuscation: generate the obfuscation candidates and obfuscate the driving statuses. The trajectory obfuscation process is described as follows.

Generate Obfuscation Candidates: The set of obfuscation candidates $\bar{\mathcal{M}} = \{\bar{m}_1, \cdots, \bar{m}_n\}$ collects all possible driving statuses of roads in the region $\mathbb{R}$, and can be calculated by the coordinator using the prior knowledge of the road network. The coordinator sends the obfuscation candidates to the vehicles once the vehicles drive into the region $\mathbb{R}$.

Obfuscate Driving Statuses: Vehicle v with the actual driving statuses $M = \{m_1, \cdots, m_i, \cdots\}$ probabilistically selects an obfuscation candidate $\bar{m}_j \in \bar{\mathcal{M}}$ as its obfuscated driving statuses $\hat{m}_i$, as given by

$$\Pr[\hat{m}_i = \bar{m}_j | m_i] = \frac{e^{-\frac{\varepsilon}{2} d(m_i, \bar{m}_j)}}{\sum\limits_{\bar{m}_x \in \bar{\mathcal{M}}} e^{-\frac{\varepsilon}{2} d(m_i, \bar{m}_x)}}, \tag{6.17}$$

where $\bar{m}_x$ is an obfuscation candidate in $\bar{\mathcal{M}}$. The obfuscation of each $m \in M$ is independent. Thus, the probability that a sequence of obfuscated driving statuses $\hat{M}$ selected based on M is given by

$$\Pr[\hat{M}|M] = \frac{e^{-\frac{\epsilon}{2}d(M,\hat{M})}}{\sum\limits_{\hat{M}_x} e^{-\frac{\epsilon}{2}d(M,\hat{M}_x)}}, \tag{6.18}$$

where $\hat{M}_x$ is any sequence of possible obfuscated driving statuses that consists of multiple $\bar{m} \in \bar{\mathcal{M}}$. The lengths of different $\hat{M}_x$ are the same.

6.4.2.3 Pseudonym Swapping

There are three steps in pseudonym swapping: the coordinator records the driving statuses, the vehicles calculate the distance between the pseudonyms, and the coordinator swaps the pseudonyms. The details are described as follows.

Record Driving Status: When the vehicles upload their driving statuses, the coordinator records the obfuscated driving statuses from each pseudonym before swapping the pseudonyms. Periodically, the coordinator records the recently uploaded driving statuses from each pseudonym that are of the same length. Then, the coordinator sends all of the recorded driving statuses and the swapping requests to the vehicles. The recorded driving statuses are deleted once the pseudonym-swapping process is completed.

Calculate Distances between Pseudonyms: One of the pseudonym swaps takes place at time T. If vehicle v_a using the pseudonym $id_a^{T\text{-}}$ needs to swap the pseudonyms, the vehicle calculates the distance between the pseudonym $id_a^{T\text{-}}$ and any pseudonym $id_x^{T\text{-}}$, denoted by $\mathcal{V}_{x,a}^{T\text{-}}$, by comparing the reported driving statuses, as given by

$$\mathcal{V}_{x,a}^{T\text{-}} = \frac{1}{2}d(\hat{M}_x^{T\text{-}}, M_a^{T\text{-}}) - \frac{1}{2}d(\hat{M}_a^{T\text{-}}, M_a^{T\text{-}}), \tag{6.19}$$

where $\hat{M}_x^{T\text{-}}$ and $id_x^{T\text{-}}$ are the obfuscated driving statuses and the pseudonym reported by any vehicle v_x before the pseudonym swap, respectively. We employ the scale factor $\frac{1}{2}$ in (6.19) to normalize $\mathcal{V}_{x,a}^{T\text{-}}$, i.e., $-0.5 \leq \mathcal{V}_{x,a}^{T\text{-}} \leq 0.5$. The calculation of $\mathcal{V}_{x,a}^{T\text{-}}$ uses the actual driving statuses of v_a and the obfuscated driving statuses of other vehicles, since v_a only knows the actual driving statuses of itself.

Swap Pseudonyms: Vehicles in $\mathcal{S}$ randomly select their new pseudonyms from the pseudonym pool $\mathcal{S}$. For a pseudonym $id_b^{T\text{-}} \in \mathcal{S}$, the probability $\Pr[id_a^{T+} = id_b^{T\text{-}}|(M_a^{T\text{-}}, id_a^{T\text{-}})]$ that v_a selects $id_b^{T\text{-}}$ as its new pseudonym id_a^{T+} is calculated by the coordinator, as given by

$$\Pr[id_a^{T+} = id_b^{T\text{-}}|(M_a^{T\text{-}}, id_a^{T\text{-}})] = \frac{e^{-\frac{\epsilon}{2}\mathcal{V}_{b,a}^{T\text{-}}}}{\sum\limits_{id_x^{T\text{-}} \in \mathcal{S}} e^{-\frac{\epsilon}{2}\mathcal{W}_{x,a}^{T\text{-}}}}, \tag{6.20}$$

where

$$\mathcal{W}_{x,a}^{T\text{-}} = \frac{1}{2}d(\hat{M}_x^{T\text{-}}, M_a^{T\text{-}}) + \frac{1}{2}d(\hat{M}_a^{T\text{-}}, M_a^{T\text{-}}). \tag{6.21}$$

A small $\mathcal{V}_{x,a}^{T\text{-}}$ indicates that $\hat{M}_x$ is close to M_a. $\mathcal{V}_{x,a}^{T\text{-}}$ is negative when $\hat{M}_x$ is closer to M_a than it is to $\hat{M}_a$. Here, v_a has a high probability of selecting $id_b^{T\text{-}}$ as its new pseudonym id_a^{T+} if $\mathcal{V}_{x,a}^{T\text{-}}$ is small. It follows from (6.20) that

$$\mathcal{V}_{x,a}^{T\text{-}} \leq \mathcal{W}_{x,a}^{T\text{-}}, \tag{6.22}$$

which leads to

$$\sum_{id_x^{T\text{-}} \in \mathcal{S}} \Pr[id_a^{T+} = id_x^{T\text{-}} | (M_a^{T\text{-}}, id_a^{T\text{-}})] \leq 1. \tag{6.23}$$

Thus, v_a may not choose any pseudonym when $\sum_{id_x^{T\text{-}} \in \mathcal{S}} \Pr[id_a^{T+} = id_x^{T\text{-}} | (M_a^{T\text{-}}, id_a^{T\text{-}})] < 1$. v_a does not change its pseudonym, i.e., $id_a^{T+} = id_a^{T\text{-}}$, if v_a does not select any id_x in $\mathcal{S}$.

If the vehicles select different pseudonyms, the coordinator returns a pseudonym to each of the vehicles. If multiple vehicles select the same pseudonym $id_x^{T\text{-}}$, the coordinator creates the same number of ring-signature-based pseudonyms using the existing works [167, 252] based on $id_x^{T\text{-}}$ and returns random new ring-signature pseudonyms to the vehicles. Hence, the vehicles can share the same pseudonym $id_x^{T\text{-}}$.

After swapping the pseudonyms at time T, the vehicles use their new pseudonyms to upload their messages with obfuscated driving statuses until the next pseudonym swap. Thus, the probability $\Pr[(\hat{M}_a^{T+}, id_a^{T+}) | (M_a^{T+}, M_a^{T\text{-}}, id_a^{T\text{-}})]$ that v_a with its actual driving statuses $M_a^{T\text{-}}, M_a^{T+}$, and pseudonym $id_a^{T\text{-}}$ generates message $(\hat{M}_a^{T+}, id_a^{T+})$ after swapping pseudonyms can be calculated as

$$\begin{aligned}
&\Pr[(\hat{M}_a^{T+}, id_a^{T+}) | (M_a^{T+}, M_a^{T\text{-}}, id_a^{T\text{-}})] \\
&= \Pr[(\hat{M}_a^{T+}, id_a^{T+}) | (M_a^{T+}, id_a^{T+})] \Pr[id_a^{T+} | (M_a^{T\text{-}}, id_a^{T\text{-}})] \\
&= \underbrace{\Pr[\hat{M}_a^{T+} | M_a^{T+}]}_{\text{Obfuscation after swap}} \sum_{\hat{M}_y^{T\text{-}}} \Big(\underbrace{\Pr[id_a^{T+} | (\hat{M}_y^{T\text{-}}, id_a^{T\text{-}})]}_{\text{Swapping pseudonyms}} \underbrace{\Pr[\hat{M}_y^{T\text{-}} | M_a^{T\text{-}}]}_{\text{Obfuscation before swap}} \Big),
\end{aligned} \tag{6.24}$$

where $M_a^{T\text{-}}$ and M_a^{T+} are the sequences of the actual driving statuses from v_a before and after swapping the pseudonyms, respectively. $\hat{M}_a^{T\text{-}}$ and $\hat{M}_a^{T+}$ can be any sequences of the obfuscated driving statuses from v_a before and after swapping the pseudonyms, respectively. $\hat{M}_y^{T\text{-}}$ is a possible sequence of the obfuscated driving statuses of v_a before swapping the pseudonyms. In JTOPS, the trajectory obfuscation is independent of the pseudonym swap, but the pseudonym swap relies on the obfuscated statuses.

6.4.3 Privacy Analysis

In this chapter, we first prove that the JTOPS with the spatial-temporal driving statuses satisfies the T-I. Then, we discuss the improvement of the distinguishability upper bounds obtained by jointly considering trajectory obfuscation and pseudonym swapping.

Theorem 6.1 *For any sequence of obfuscated messages* $(\hat{M}_x^{T+}, id_x^{T+})$*, the JTOPS mechanism satisfies the* $(\varepsilon, \mathbb{R})$*-Trajectory-Indistinguishability that any two vehicles* v_a *and* v_b *in region* $\mathbb{R}$ *are* $(\varepsilon, \mathbb{R})$*-trajectory-indistinguishable.*

Proof 6.2 *Assuming that the pseudonyms are swapped at time* T*, we can substitute* (6.17)*,* (6.19)*, and* (6.20) *into* (6.24)*, and evaluate the distinguishability of* v_a *and* v_b*, i.e., the difference of the probabilities that* v_a *with* $(M_a^{T+}, M_a^{T\text{-}}, id_a^{T\text{-}})$ *and* v_b *with* $(M_b^{T+}, M_b^{T\text{-}}, id_b^{T\text{-}})$ *generate* $(\hat{M}_x^{T+}, id_x^{T+})$*, as given in* (6.25).

$$\frac{\Pr[(\hat{M}_x^{T+}, id_x^{T+})|(M_a^{T+}, M_a^{T\text{-}}, id_a^{T\text{-}})]}{\Pr[(\hat{M}_x^{T+}, id_x^{T+})|(M_b^{T+}, M_b^{T\text{-}}, id_b^{T\text{-}})]}$$
$$= \underbrace{\frac{\dfrac{e^{-\frac{\epsilon}{2}d(\hat{M}_x, M_a^{T+})}}{\sum_{\hat{M}_z} e^{-\frac{\epsilon}{2}d(\hat{M}_z, M_a^{T+})}}}{\dfrac{e^{-\frac{\epsilon}{2}d(\hat{M}_x, M_b^{T+})}}{\sum_{\hat{M}_z} e^{-\frac{\epsilon}{2}d(\hat{M}_z, M_b^{T+})}}}}_{\text{Obfuscation after swap}}$$
$$\times \frac{\sum_{\hat{M}_y^{T\text{-}}} \left(\dfrac{e^{-\frac{\epsilon}{2}\mathcal{V}_{x,a}^{T\text{-}}}}{\sum_{id_c^{T\text{-}} \in \mathcal{S}} e^{-\frac{\epsilon}{2}\mathcal{W}_{c,a}^{T\text{-}}}} \times \dfrac{e^{-\frac{\epsilon}{2}d(\hat{M}_y^{T\text{-}}, M_a^{T\text{-}})}}{\sum_{\hat{M}_z} e^{-\frac{\epsilon}{2}d(\hat{M}_z^{T\text{-}}, M_a^{T\text{-}})}} \right)}{\sum_{\hat{M}_y^{T\text{-}}} \left(\underbrace{\dfrac{e^{-\frac{\epsilon}{2}\mathcal{V}_{x,b}^{T\text{-}}}}{\sum_{id_c^{T\text{-}} \in \mathcal{S}} e^{-\frac{\epsilon}{2}\mathcal{W}_{c,b}^{T\text{-}}}}}_{\text{Swapping Pseudonyms}} \times \underbrace{\dfrac{e^{-\frac{\epsilon}{2}d(\hat{M}_y^{T\text{-}}, M_b^{T\text{-}})}}{\sum_{\hat{M}_z} e^{-\frac{\epsilon}{2}d(\hat{M}_z^{T\text{-}}, M_b^{T\text{-}})}}}_{\text{Obfuscation before swap}} \right)}, \tag{6.25}$$

In (6.25), $\hat{M}_z$, consisting of the obfuscation candidates $\bar{m} \in \bar{\mathcal{M}}$, is a possible sequence of the driving statuses in region $\mathbb{R}$. $\hat{M}_x^{T\text{-}}$ is the sequence of obfuscated driving statuses reported with $id_x^{T\text{-}}$. $(\hat{M}_c^{T\text{-}}, id_c^{T\text{-}})$ is the sequence of the uploaded messages reported by any other vehicle $v_c \in \mathcal{S}$. $\hat{M}_y^{T\text{-}}$ is a possible sequence of the driving statuses of v_a and v_b before time T.

To simplify (6.25), we let

$$f(M_j) = \sum_{\hat{M}_z} e^{-\frac{\epsilon}{2}d(\hat{M}_z, M_j)}; \tag{6.26}$$

$$h(M_j) = \sum_{id_c^{T\text{-}} \in \mathcal{S}} e^{-\frac{\epsilon}{2}\mathcal{W}_{c,j}^{T\text{-}}}; \tag{6.27}$$

$$\mathcal{H}(M_j, \hat{M}_y^{T\text{-}}) = \frac{1}{h(M_j)} e^{-\frac{\epsilon}{2}\mathcal{V}_{x,a}^{T\text{-}}} \times \frac{1}{f(M_j^{T\text{-}})} e^{-\frac{\epsilon}{2}d(\hat{M}_y^{T\text{-}}, M_j^{T\text{-}})}. \tag{6.28}$$

Thus, (6.25) can be rewritten as

$$\frac{\Pr[(\hat{M}_x^{T+}, id_x^{T+})|(M_a^{T+}, M_a^{T\text{-}}, id_a^{T\text{-}})]}{\Pr[(\hat{M}_x^{T+}, id_x^{T+})|(M_b^{T+}, M_b^{T\text{-}}, id_b^{T\text{-}})]} = \frac{f(M_b^{T+})}{f(M_a^{T+})} \times e^{\frac{\epsilon}{2}\left(d(\hat{M}_x^{T+}, M_b^{T+}) - d(\hat{M}_x^{T+}, M_a^{T+})\right)} \times \frac{\sum_{\hat{M}_y^{T\text{-}}} \mathcal{H}(M_a, \hat{M}_y^{T\text{-}})}{\sum_{\hat{M}_y^{T\text{-}}} \mathcal{H}(M_b, \hat{M}_y^{T\text{-}})}. \tag{6.29}$$

For any obfuscated driving statuses $(\hat{M}_x^{T\text{-}}, \hat{M}_x^{T+})$ from v_x, as the driving statuses hold the triangle inequality, we have

$$\mathcal{D}(v_b, v_x) - \mathcal{D}(v_a, v_x) \leq \mathcal{D}(v_a, v_b). \tag{6.30}$$

Thus, (6.29) meets the following inequality

$$\frac{\Pr[(\hat{M}_x^{T+}, id_x^{T+})|(M_a^{T+}, M_a^{T\text{-}}, id_a^{T\text{-}})]}{\Pr[(\hat{M}_x^{T+}, id_x^{T+})|(M_b^{T+}, M_b^{T\text{-}}, id_b^{T\text{-}})]} \leq \frac{f(M_b^{T+})}{f(M_a^{T+})} \times e^{\frac{\varepsilon}{2}d(M_a^{T+},M_b^{T+})} \times \frac{\sum_{\hat{M}_y^{T\text{-}}} \mathcal{H}(M_a, \hat{M}_y^{T\text{-}})}{\sum_{\hat{M}_y^{T\text{-}}} \mathcal{H}(M_b, \hat{M}_y^{T\text{-}})}. \tag{6.31}$$

By employing the triangle inequality, i.e., $d(\hat{M}_z, M_b^{T+}) \geq d(\hat{M}_z, M_a^{T+}) - d(M_a^{T+}, M_b^{T+})$, *we have*

$$e^{-\frac{\varepsilon}{2}d(\hat{M}_z,M_b^{T+})} \leq e^{-\frac{\varepsilon}{2}\left(d(\hat{M}_z,M_a^{T+})-d(M_a^{T+},M_b^{T+})\right)}. \tag{6.32}$$

Based on (6.32), *for all* $\hat{M}_z$ *consisting of* $\bar{m}$, *the following holds*

$$\sum_{\hat{M}_z}\left(e^{-\frac{\varepsilon}{2}d(\hat{M}_z,M_b^{T+})} - e^{-\frac{\varepsilon}{2}\left(d(\hat{M}_z,M_a^{T+})-d(M_a^{T+},M_b^{T+})\right)}\right) \leq 0,$$

which can be rewritten as

$$\sum\nolimits_{\hat{M}_z} e^{-\frac{\varepsilon}{2}d(\hat{M}_z,M_b^{T+})} - e^{\frac{\varepsilon}{2}d(M_a^{T+},M_b^{T+})}\sum\nolimits_{\hat{M}} e^{-\frac{\varepsilon}{2}d(\hat{M}_z,M_a^{T+})} \leq 0. \tag{6.33}$$

Based on the definition of $f(M_j)$ *in* (6.26), *we have*

$$f(M_b^{T+}) - e^{\frac{\varepsilon}{2}d(M_a^{T+},M_b^{T+})} f(M_a^{T+}) \leq 0, \tag{6.34}$$

which leads to

$$\frac{f(M_b^{T+})}{f(M_a^{T+})} \leq e^{\frac{\varepsilon}{2}d(M_a^{T+},M_b^{T+})}. \tag{6.35}$$

By multiplying $e^{\frac{\varepsilon}{2}d(M_a^{T+},M_b^{T+})}$ *on the both sides of* (6.35), *it follows that*

$$\frac{f(M_b^{T+})}{f(M_a^{T+})} \times e^{\frac{\varepsilon}{2}d(M_a^{T+},M_b^{T+})} \leq e^{\varepsilon d(M_a^{T+},M_b^{T+})}. \tag{6.36}$$

Combining (6.31) *and* (6.36), *we have*

$$\frac{\Pr[(\hat{M}_x^{T+}, id_x^{T+})|(M_a^{T+}, M_a^{T\text{-}}, id_a^{T\text{-}})]}{\Pr[(\hat{M}_x^{T+}, id_x^{T+})|(M_b^{T+}, M_b^{T\text{-}}, id_b^{T\text{-}})]} \leq e^{\varepsilon d(M_a^{T+},M_b^{T+})} \times \frac{\sum_{\hat{M}_y^{T\text{-}}} \mathcal{H}(M_a, \hat{M}_y^{T\text{-}})}{\sum_{\hat{M}_y^{T\text{-}}} \mathcal{H}(M_b, \hat{M}_y^{T\text{-}})}. \tag{6.37}$$

Based on the definition of $\mathcal{H}(M_j, \hat{M}_y^{T\text{-}})$ *in* (6.28), *we have*

$$\frac{\mathcal{H}(M_a, \hat{M}_y^{T\text{-}})}{\mathcal{H}(M_b, \hat{M}_y^{T\text{-}})} = \frac{h(M_b)}{h(M_a)} \times \frac{f(M_b^{T\text{-}})}{f(M_a^{T\text{-}})} \times e^{\frac{\varepsilon}{2}(\mathcal{V}_{x,b}^{T\text{-}} - \mathcal{V}_{x,a}^{T\text{-}})} \times e^{\frac{\varepsilon}{2}\left(d(\hat{M}_y^{T\text{-}},M_b^{T\text{-}})-d(\hat{M}_y^{T\text{-}},M_a^{T\text{-}})\right)}, \tag{6.38}$$

By following the analysis from (6.32) to (6.36), we attain

$$\frac{f(M_b^{T\text{-}})}{f(M_a^{T\text{-}})} \leq e^{\frac{\varepsilon}{2} d(M_a^{T\text{-}}, M_b^{T\text{-}})}. \tag{6.39}$$

Based on the triangle inequality of driving status, we have

$$\frac{e^{\frac{\varepsilon}{2}\left(\frac{1}{2}d(\hat{M}_c^{T\text{-}}, M_b^{T\text{-}}) - \frac{1}{2}d(\hat{M}_b^{T\text{-}}, M_b^{T\text{-}})\right)}}{e^{\frac{\varepsilon}{2}\left(\frac{1}{2}d(\hat{M}_c^{T\text{-}}, M_a^{T\text{-}}) - \frac{1}{2}d(\hat{M}_a^{T\text{-}}, M_a^{T\text{-}})\right)}} \leq e^{\frac{\varepsilon}{2}\left(\frac{1}{2}d(M_a^{T\text{-}}, M_b^{T\text{-}}) - \frac{1}{2}d(\hat{M}_b^{T\text{-}}, M_b^{T\text{-}}) + \frac{1}{2}d(\hat{M}_a^{T\text{-}}, M_a^{T\text{-}})\right)}. \tag{6.40}$$

As $0 \leq d(M_b^{T\text{-}}, M_a^{T\text{-}}) \leq 1$, *we have*

$$\begin{aligned} & e^{-\frac{\varepsilon}{2}\left(\frac{1}{2}d(\hat{M}_c^{T\text{-}}, M_b^{T\text{-}}) + \frac{1}{2}d(\hat{M}_b^{T\text{-}}, M_b^{T\text{-}})\right)} \\ & \leq e^{-\frac{\varepsilon}{2}\left(\frac{1}{2}d(\hat{M}_c^{T\text{-}}, M_a^{T\text{-}}) - \frac{1}{2}d(M_b^{T\text{-}}, M_a^{T\text{-}}) + \frac{1}{2}d(\hat{M}_b^{T\text{-}}, M_b^{T\text{-}})\right)}. \end{aligned} \tag{6.41}$$

By following the analysis in (6.32) - (6.36), we have

$$\begin{aligned} & \frac{\sum_{id_c^{T\text{-}} \in \mathcal{S}} e^{-\frac{\varepsilon}{2}\left(\frac{1}{2}d(\hat{M}_c^{T\text{-}}, M_b^{T\text{-}}) + \frac{1}{2}d(\hat{M}_b^{T\text{-}}, M_b^{T\text{-}})\right)}}{\sum_{id_c^{T\text{-}} \in \mathcal{S}} e^{-\frac{\varepsilon}{2}\left(\frac{1}{2}d(\hat{M}_c^{T\text{-}}, M_a^{T\text{-}}) + \frac{1}{2}d(\hat{M}_a^{T\text{-}}, M_a^{T\text{-}})\right)}} \\ & \leq e^{\frac{\varepsilon}{2}\left(\frac{1}{2}d(M_b^{T\text{-}}, M_a^{T\text{-}}) + \frac{1}{2}d(\hat{M}_a^{T\text{-}}, M_a^{T\text{-}}) - \frac{1}{2}d(\hat{M}_b^{T\text{-}}, M_b^{T\text{-}})\right)}. \end{aligned} \tag{6.42}$$

Based on the definition of $h(M_j)$ *in (6.27), it follows:*

$$\frac{h(M_b)}{h(M_a)} \leq e^{\frac{\varepsilon}{2}\left(\frac{1}{2}d(M_a^{T\text{-}}, M_b^{T\text{-}}) - \frac{1}{2}d(\hat{M}_b^{T\text{-}}, M_b^{T\text{-}}) + \frac{1}{2}d(\hat{M}_a^{T\text{-}}, M_a^{T\text{-}})\right)}. \tag{6.43}$$

Combining (6.40) and (6.43), we have

$$\frac{h(M_b)}{h(M_a)} \times \frac{e^{\frac{\varepsilon}{2}\mathcal{V}_{x,b}^{T\text{-}}}}{e^{\frac{\varepsilon}{2}\mathcal{V}_{x,a}^{T\text{-}}}} \leq e^{\frac{\varepsilon}{2}\left(d(M_a^{T\text{-}}, M_b^{T\text{-}}) - d(\hat{M}_b^{T\text{-}}, M_b^{T\text{-}}) + d(\hat{M}_a^{T\text{-}}, M_a^{T\text{-}})\right)}, \tag{6.44}$$

which leads to

$$\frac{\mathcal{H}(M_a^{T\text{-}}, \hat{M}_y^{T\text{-}})}{\mathcal{H}(M_b^{T\text{-}}, \hat{M}_y^{T\text{-}})} \leq e^{\varepsilon d(M_a^{T\text{-}}, M_b^{T\text{-}})}. \tag{6.45}$$

As $\hat{M}_y^{T\text{-}}$ *can be any possible obfuscated driving status, the following holds*

$$\frac{\sum\limits_{\hat{M}_y^{T\text{-}}} \mathcal{H}(M_a^{T\text{-}}, \hat{M}_y^{T\text{-}})}{\sum\limits_{\hat{M}_y^{T\text{-}}} \mathcal{H}(M_b^{T\text{-}}, \hat{M}_y^{T\text{-}})} \leq e^{\varepsilon d(M_a^{T\text{-}}, M_b^{T\text{-}})}. \tag{6.46}$$

By substituting (6.46) into (6.37), we finally have

$$\begin{aligned} \frac{\Pr[(\hat{M}_x^{T+}, id_x^{T+}) | (M_a^{T+}, M_a^{T\text{-}}, id_a^{T\text{-}})]}{\Pr[(\hat{M}_x^{T+}, id_x^{T+}) | (M_b^{T+}, M_b^{T\text{-}}, id_b^{T\text{-}})]} & \leq e^{\varepsilon d(M_a^{T\text{-}}, M_b^{T\text{-}})} e^{\varepsilon d(M_a^{T+}, M_b^{T+})} \\ & = e^{\varepsilon \mathcal{D}(v_a, v_b)}, \end{aligned} \tag{6.47}$$

which satisfies (6.16). In other words, for any sequence of uploaded messages $(\hat{M}_x^{T+}, id_x^{T+})$, *vehicles* v_a *and* v_b *in region* $\mathbb{R}$ *are* $(\varepsilon, \mathbb{R})$*-trajectory-indistinguishable. Hence, the introduced JTOPS mechanism can protect privacy under the external GPA.*

We find that pseudonym swapping can be considered a differential privacy process.

Corollary 6.2 *Pseudonym swapping can be treated as a differential privacy process.*

Proof 6.3 *Let id' be any pseudonym after pseudonym swapping, and $\{id_1, id_2, \cdots, id_n\}$ be the identity set that can be swapped to id'. For any id_i and id_j ($i, j \in [1, n]$), the following holds,*

$$\begin{cases} 0 < \Pr[id'|id_i] \leq 1, \\ 0 < \Pr[id'|id_j] \leq 1, \end{cases}$$

Let $\max(\frac{\Pr[id'|id_i]}{\Pr[id'|id_j]}) = x$. *According to the definition of* $\Pr[id'|id_i]$ *and* $\Pr[id'|id_j]$, *it is easy to prove* $x \geq 1$.

Thus we have

$$\frac{1}{x} \leq \frac{\Pr[id'|id_i]}{\Pr[id'|id_j]} \leq x. \tag{6.48}$$

Let $\varepsilon = \ln x$. *The above equation can be transformed to*

$$e^{-\varepsilon} \leq \frac{\Pr[id'|id_i]}{\Pr[id'|id_j]} \leq e^{\varepsilon}, \tag{6.49}$$

which is an expression of differential privacy.

We next prove the superiority of the JTOPS, compared to the pseudonym-swapping-only or trajectory obfuscation-only approach.

Corollary 6.3 *Applying pseudonym swapping and trajectory obfuscation jointly can provide higher location privacy-preserving capability than using either pseudonym swapping or trajectory obfuscation.*

Proof 6.4 *When using pseudonym swapping and trajectory obfuscation separately, the pseudonyms are swapped based on the actual driving statuses. For any two vehicles v_a and v_b at time T, the mechanism, which separately swaps the pseudonyms and obfuscates the driving statuses, can be written as*

$$\frac{\Pr[(\hat{M}_a^{T\text{-}}, id_a^{T+})|(M_a^{T\text{-}}, id_a^{T\text{-}})]}{\Pr[(\hat{M}_b^{T\text{-}}, id_b^{T+})|(M_b^{T\text{-}}, id_b^{T\text{-}})]} = \underbrace{\frac{\Pr[\hat{M}_a^{T\text{-}}|M_a^{T\text{-}}]}{\Pr[\hat{M}_b^{T\text{-}}|M_b^{T\text{-}}]}}_{\textit{Obfuscation before swap}} \times \underbrace{\frac{\Pr[id_a^{T+}|id_a^{T\text{-}}]}{\Pr[id_b^{T+}|id_b^{T\text{-}}]}}_{\textit{Swapping pseudonym}}. \tag{6.50}$$

Assuming that the separate use of the pseudonym swapping and trajectory obfuscation is based on DP, according to the definition of $d(M_a^{T+}, v_b^{T+})$ and corollary 1, we have

$$\begin{cases} \dfrac{\Pr[\hat{M}_a^{T\text{-}}|M_a^{T\text{-}}]}{\Pr[\hat{M}_b^{T\text{-}}|M_b^{T\text{-}}]} \leq e^{\varepsilon d(M_a^{T\text{-}}, M_b^{T\text{-}})}; \\ \dfrac{\Pr[id_a^{T+}|id_a^{T\text{-}}]}{\Pr[id_b^{T+}|id_b^{T\text{-}}]} \leq e^{\varepsilon'}, \end{cases}$$

where ε' is the privacy budget of the pseudonym-swapping process. The above equations lead to

$$\frac{\Pr[(\hat{M}_a^{T+}, id_a^{T+})|(M_a^{T+}, M_a^{T-}; id_a^{T-})]}{\Pr[(\hat{M}_b^{T+}, id_b^{T+})|(M_b^{T+}, M_b^{T-}; id_b^{T-})]} \leq e^{\varepsilon d(M_a^{T-}, M_b^{T-}) + \varepsilon'}. \tag{6.51}$$

In the same case, according to the definition of $\mathcal{H}(M_j, \hat{M}_y^{T-})$ in (6.28) and (6.32) - (6.45), the JTOPS mechanism can be written as

$$\frac{\Pr[(\hat{M}_a^{T-}; id_a^{T+})|(M_a^{T-}; id_a^{T-})]}{\Pr[(\hat{M}_b^{T-}; id_b^{T+})|(M_b^{T-}; id_b^{T-})]} = \frac{\mathcal{H}(M_a^{T-}; \hat{M}_y^{T-})}{\mathcal{H}(M_b^{T-}; \hat{M}_y^{T-})} \leq e^{\varepsilon d(M_a^{T-}, M_b^{T-})}. \tag{6.52}$$

By comparing the right-hand sides of (6.51) and (6.52), jointly utilizing the trajectory obfuscation and pseudonym swapping has a lower upper bound of indistinguishability, indicating the JTOPS achieves higher privacy protection than using trajectory obfuscation or pseudonym swapping only.

The JTOPS swaps the pseudonyms based on the obfuscated driving statuses of the vehicles, which does not increase the upper bound of the indistinguishability.

6.4.4 Under Collusion Attack

An adversary can potentially collude with the coordinator, who is aware of the pseudonym-swapping results. Then, the old and new pseudonyms can be precisely linked. In this chapter, we analyze how the JTOPS can limit the coordinator's knowledge to protect the vehicles' location privacy. Compared with the existing pseudonym-based mechanisms [224, 51], our mechanism can still protect location privacy in the presence of an untrusted or collusive coordinator.

If an adversary can precisely link the new and old pseudonyms, the estimation probability (EP) that the adversary can accurately estimate the actual driving status of the vehicle is defined as follows.

Definition 6.2 (EP) *Given a set of obfuscated driving statuses $\hat{M} = \{\hat{m}_1, \cdots, \hat{m}_i, \cdots\}$ from vehicle v with the actual driving statuses $M = \{m_1, \cdots, m_i, \cdots\}$, the EP of the vehicle i.e., $\theta(v)$, is given by*

$$\theta(v) = \frac{1}{k} \sum\nolimits_{i=1}^{k} \Pr[m_i|\hat{m}_i], \tag{6.53}$$

where $\hat{m}_i$ is the obfuscated driving status of vehicle v based on the i-th actual m_i. $\Pr[m_i|\hat{m}_i]$ is based on Bayes' theorem [255]. A low EP indicates a high capability for privacy preservation.

As the pseudonym-swapping results are known by the adversary, we introduce our previous work RN-I [255] to show the JTOPS can still provide location privacy. By extending the metrics, the RN-I in this chapter is defined as follows.

Definition 6.3 (RN-I) *A mechanism satisfies $(\varepsilon, \mathbb{R})$-RN-I if and only if, with obfuscated driving statuses $\hat{M}_a$, any actual driving status pair (M_{a_1}, M_{a_2}) of v_a yield*

$$\frac{\Pr[\hat{M}_a|M_{a_1}]}{\Pr[\hat{M}_a|M_{a_2}]} \leq e^{\varepsilon d(M_{a_1}, M_{a_2})}. \tag{6.54}$$

Theorem 6.2 *If an adversary colludes with the coordinator, the JTOPS mechanism can protect the privacy of vehicles, such that, for the obfuscated driving statuses $\hat{M}_a$, any set of driving statuses M_{a_1} and M_{a_2} of vehicle v_a satisfies RN-I.*

Proof 6.5 *We assume that pseudonym-swapping results are exposed. Thus, the JTOPS would be written as*

$$\frac{\Pr[\hat{M}_a|M_{a_1}]}{\Pr[\hat{M}_a|M_{a_2}]}, \tag{6.55}$$

where $\hat{M}_a$ is the reported trajectory of vehicle v_a.

By substituting (6.17) *into* (6.55), *the latter can be written as*

$$\frac{\Pr[\hat{M}_a|M_{a_1}]}{\Pr[\hat{M}_a|M_{a_2}]} = \frac{\frac{e^{-\frac{\varepsilon}{2}d(M_{a_1},\hat{M}_a)}}{\sum_{\hat{M}_{a_x}} e^{-\frac{\varepsilon}{2}d(M_{a_1},\hat{M}_{a_x})}}}{\frac{e^{-\frac{\varepsilon}{2}d(M_{a_2},\hat{M}_a)}}{\sum_{\hat{M}_{a_x}} e^{-\frac{\varepsilon}{2}d(M_{a_2},\hat{M}_{a_x})}}}, \tag{6.56}$$

where $\hat{M}_{a_x}$ is any sequence of obfuscated driving statuses that can be generated by vehicle v_a.

By following the analysis from (6.32) *to* (6.36), *we have*

$$d(\hat{M}_a, M_{a_2}) - d(\hat{M}_a, M_{a_1}) \leq d(M_{a_1}, M_{a_2}). \tag{6.57}$$

Let $g(M_{a_i}) = \sum_{\hat{M}_x} e^{-\frac{\varepsilon}{2}d(M_{a_i},\hat{M}_x)}$, we have

$$\frac{\Pr[\hat{M}_a|M_{a_1}]}{\Pr[\hat{M}_a|M_{a_2}]} = \frac{g(M_{a_2})}{g(M_{a_1})} e^{\frac{\varepsilon}{2}\left(d(M_{a_2},\hat{M}_a)-d(M_{a_1},\hat{M}_a)\right)} \leq \frac{g(M_{a_2})}{g(M_{a_1})} e^{\frac{\varepsilon}{2}d(M_{a_1},M_{a_2})}. \tag{6.58}$$

By employing the triangle inequality, we have

$$e^{-\frac{\varepsilon}{2}d(M_{a_2},\hat{M}_x)} \leq e^{-\frac{\varepsilon}{2}\left(d(M_{a_1},\hat{M}_x)-d(M_{a_1},M_{a_2})\right)}. \tag{6.59}$$

Following the analysis of (6.32)–(6.36), *we have*

$$\frac{g(M_{a_2})}{g(M_{a_1})} \leq e^{\frac{\varepsilon}{2}d(M_{a_1},M_{a_2})}. \tag{6.60}$$

By substituting (6.60) *into* (6.58), *the latter can be rewritten as*

$$\frac{\Pr[\hat{M}_a|M_{a_1}]}{\Pr[\hat{M}_a|M_{a_2}]} \leq e^{\varepsilon d(M_{a_1},M_{a_2})}, \tag{6.61}$$

which satisfies the RN-I. In other words, for any sequence $\hat{M}_a$ of obfuscated driving statuses from vehicle v_a, any two historical driving status sequences M_{a_1} and M_{a_2} in the prior knowledge are $(\varepsilon, \mathbb{R})$-Road Network-indistinguishable. Hence, the introduced JTOPS mechanism can resist internal GPAs.

6.5 Experimental Results

In this chapter, we compare the JTOPS with the existing hybrid mechanism [402], which is also based on pseudonyms and obfuscation in the same scenario as considered in this chapter. We also compare the trajectory obfuscation process and pseudonym-swapping process of the JTOPS with the two latest mechanisms developed in [255] and [224].

The experiments are conducted with two real-world road networks[2] and the T-Drive trajectory (10,357 drivers in Beijing, China) [462]. The two real-world road networks are extracted from the Open Street Map.[3] We utilize the trajectories of vehicles, which have at least ten trajectory points in the experimental road networks. We set $\alpha_1 = \alpha_2 = \alpha_3 = \alpha_4 = 0.25$. In the experiments, we average the ASR (6.11), DU (6.12), PU (6.13), and EP (6.53) of multiple vehicles by running the mechanisms locally on the vehicles for 50 times.

6.5.1 Adversary's Success Rate and Estimation Probability

The comparison of the ASR (6.11) among the JTOPS, Multilevel Location Privacy Scheme (MLPS) [402], and Pseudonym Swap with Provable Unlinkability (PAPU) [224] mechanism is shown in Figure 6.2(a). The ASRs of PAPU and MLPS are higher than 80%, and do not change with ε and the number of vehicles in $\mathcal{S}$. The reason is that the two mechanisms require the vehicles to report actual driving statuses. Adversaries can infer the actual identity of each vehicle and ignore pseudonym swaps by matching the obtained driving statuses with their prior knowledge of the vehicles. The ASR of the PAPU mechanism is higher than that of the MLPS in both large and small regions, as the MLPM randomly selects cooperative vehicles, and the PAPU selects candidates with pseudonym-swapping probabilities. Compared with the two mechanisms, the introduced mechanism achieves a low ASR in both two road networks. The JTOPS allows the vehicles to upload obfuscated driving statuses so that the adversary cannot gain any useful information by eavesdropping. Even if the adversary has prior knowledge of the vehicles, the driving statuses can still be protected by the introduced JTOPS.

The ASRs of the JTOPS with different numbers of vehicles, ε, and region sizes are shown in Figure 6.2(b). The ASR increases with ε. It is evident that the JTOPS can provide better privacy protection with a smaller ε. In the large region size, the ASR of the JTOPS is approximately 21% and 20.6% when ε is 10 and 5 vehicles. When 25 vehicles swap their pseudonyms, the ASR decreases to 5% in both $\varepsilon = 5$ and $\varepsilon = 10$. The adversary cannot gain useful knowledge by analyzing the obtained driving statuses of the vehicles as the uploaded driving statuses are obfuscated. As a result, the adversary cannot match their prior knowledge, and it can only randomly guess the actual identity of the vehicle. By randomly guessing the actual identity, the number of vehicles in $\mathcal{S}$ has a stronger impact on the ASR than ϵ.

The comparison of the EP (6.53) between the JTOPS and Personalized Location Privacy-Preserving mechanism (PLPP) [255] is shown in Figure 6.2(c). The PLPP mechanism takes the route distance as the only metric, while the JTOPS extends the metric to the location distance, direction, speed, and driving time of the vehicles. As shown in

[2]There are 3,246 trajectory points of 119 vehicles, 34 connections and 63 roads within the small region size (116.3595°$\leq$ longitude $\leq$ 116.3645°and 39.9135°$\leq$ latitude $\leq$ 39.9085°). There are 20,745 trajectory points of 558 vehicles, 181 connections, and 386 roads within the large region size (116.357°$\leq$ longitude $\leq$ 116.367°and 39.906°$\leq$ latitude $\leq$ 39.916°).

[3]Open Street Map is an open-source database of the world's geographic map. `https://www.openstreetmap.org/`

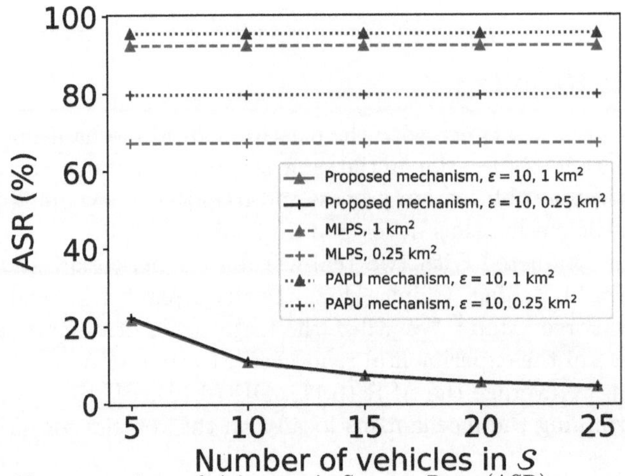

(a) The comparison of Adversary's Success Rate (ASR) among our JTOPS, MLPS [402], and the PAPU mechanism [224].

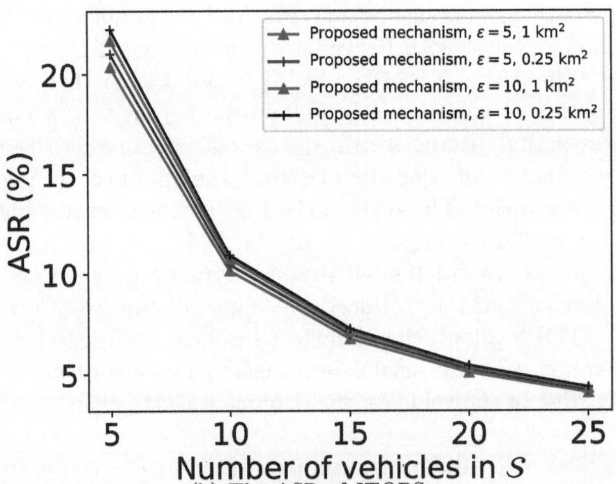

(b) The ASR of JTOPS.

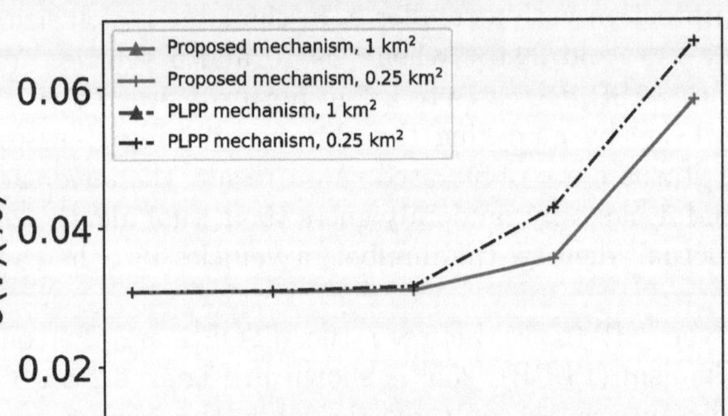

(c) The comparison of EP between the JTOPS and PLPP mechanism [255].

FIGURE 6.2: The comparison of privacy protection and data utility.

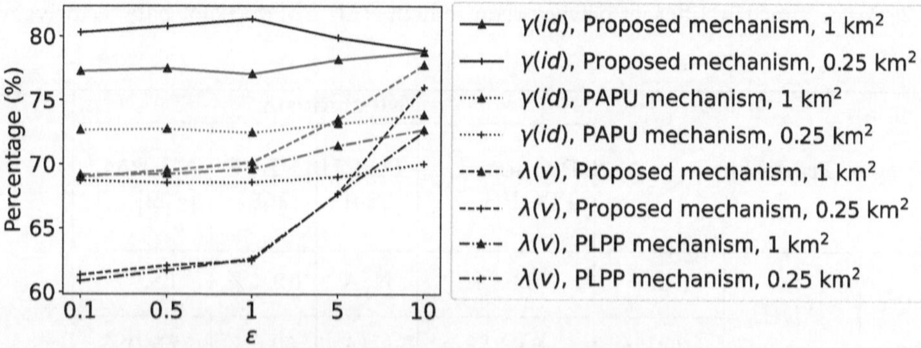

FIGURE 6.3: The comparison of data and pseudonym utility.

Figure 6.2(c), the introduced mechanism provides a higher privacy-preserving capability than the PLPP mechanism in the small region. The distance between the two mechanisms is significant under a high ε value. In the large region, the two mechanisms have similar estimation probabilities of less than 0.01%.

6.5.2 Data Utility and Pseudonym Utility

We proceed to compare the trajectory obfuscation process and the pseudonym-swapping process of the JTOPS with the PLPP and PAPU mechanisms in Figure 6.3. As shown in Figure 6.3, the trajectory obfuscation process of the JTOPS, which considers multiple driving statuses, achieves a higher data utility than the PLPP mechanism. The improvement of data utility is significant under a high ε because the mechanism has a high probability of selecting the candidate close to the actual vehicle. The ADU (6.12) of the JTOPS is higher in data utility in the large regions than it is in the small regions. This is because the obfuscation candidate set is fine-grained in the large region, as the densities of the two regions are similar. Thus, the vehicles have more choices, which are close to their actual driving statuses, in the large region than they have in the small region. The difference in the DU between the JTOPS and PLPP is smaller in the small regions than it is in the large regions because the driving statuses between the vehicles and candidates are close in a small region.

As shown in Figure 6.3, The pseudonyms have a high data utility by using the JTOPS, and the driving statuses reported with the pseudonym before the swapping process are similar to those of the pseudonym after the swapping process. The reason is that the actual driving statuses of the vehicles in road networks have high randomness, indicating the distance in the actual driving statuses among various vehicles could be significant. The introduced mechanism utilizes obfuscated driving statuses that are selected from the obfuscated candidates. Thus, the distance in the obfuscated driving statuses is stable in fixed road networks. As the large region and the small region have a similar road density, the driving statuses of the candidates in the small region are closer to each other than those in the large region. For the PAPU mechanism, the driving status granularity of the vehicles in the small region is coarser than those in the large region, so the PU of the PAPU mechanism in the small region is less than it is in the large region.

TABLE 6.2: Performance in different privacy requirements (MLPM does not balance privacy and utility).

Trade-off			Mechanism			
			Proposed JTOPS	**PAPU [224]**	**PLPP [255]**	**MLPM [402]**
High Privacy ($\varepsilon = 0.1$)	ADU	1 km^2	68.8%	N/A	**69.1%**	68.2%
		0.25 km^2	**61.3%**	N/A	60.6%	**75.9%**
	PU	1 km^2	**77.2%**	72.5%	N/A	N/A
		0.25 km^2	**80.4%**	68.9%	N/A	N/A
	ASR (10 vehicles in $\mathcal{S}$)	1 km^2	**10.0%**	93.2%	N/A	91.6%
		0.25 km^2	**10.2%**	80.1%	N/A	77.2%
High Utility ($\varepsilon = 100$)	ADU	1 km^2	**77.7%**	N/A	72.3%	68.2%
		0.25 km^2	75.8%	N/A	72.4%	**75.9%**
	PU	1 km^2	**79.3%**	73.9%	N/A	N/A
		0.25 km^2	**79.1%**	70.4%	N/A	N/A
	ASR (10 vehicles in $\mathcal{S}$)	1 km^2	**11.1%**	93.2%	N/A	91.6%
		0.25 km^2	**11.9%**	80.1%	N/A	72.2%

6.5.3 Performance in Different Privacy Requirements

The comparisons of ADU, PU, and ASR under different privacy requirements among the JTOPS, PAPU [224], PLPP [255], MLPM [402] are shown in Table 6.2. We follow the privacy setting of [255], where $\varepsilon = 0.1$ represents that the drivers need a high level of privacy protection and $\varepsilon = 10$ represents that drivers need a high level of data and pseudonym utility. A high ADU and PU correspond to a high utility of data and pseudonyms, while a low ASR means high-privacy protection.

In the ASR comparison, we assume there are ten vehicles in the pseudonym-swapping process. Compared with PLPP, the ASRs of PAPU and MLPM are high, and the values are around 80% in the small region and more than 90% in the large regions. This is because the two mechanisms do not protect the uploaded messages so that the adversary can accurately match the pseudonyms with the actual identities based on its prior knowledge. The ASRs of PAPU and MLPM do not change with ε because ε does not influence the contents of the uploaded messages. The JTOPS achieves a significant location privacy-preserving capability that its ASR values are almost 10% in each case.

In the utility comparison, the ADUs of the JTOPS and PLPP are significantly higher in the high-utility requirement case than they are in the high-privacy requirement case.

This is because the two mechanisms have a high probability of selecting candidates who are similar to the actual driving statuses. The ADU of MLPM is lower than those of the two mechanisms because MLPM selects cooperators randomly. The ADU of PAPU is not considered in this comparison, as PAPU does not protect actual driving statuses. The PUs of the JTOPS and PAPU are similar, but the PU of JTOPS is a bit higher than that of PAPU. One reason is that the JTOPS addresses the conflict of pseudonym swapping, which is overlooked by PAPU. Another reason is that the pseudonym-swapping probability functions of the two mechanisms are different, where the JTOPS allows the vehicles to select a similar candidate with a high probability. Nevertheless, the JTOPS still satisfies the concept of DP, which is explained below.

Let the actual driving status sequence of vehicle v is $M = \{m_1, \cdots, m_l\}$, and the obfuscation candidate set is $\bar{\mathcal{M}} = \{\bar{m}_1, \cdots, \bar{m}_n\}$. The obfuscated driving statuses sequence of v is $\hat{M} = \{\hat{m}_1, \cdots, \hat{m}_l\}$. As each obfuscated driving status $\hat{m}$ is independently selected, according to (6.17), we have

$$\Pr[\hat{M}|M] = \prod_{m_i \in M} \frac{e^{-\frac{\varepsilon}{2}d(m_i,\hat{m}_i)}}{\sum_{\bar{m}_x \in \bar{\mathcal{M}}} e^{-\frac{\varepsilon}{2}d(m_i,\bar{m}_x)}}. \tag{6.62}$$

For any $\hat{m}_i \in \hat{M}$, $\hat{m}$ is selected from $\bar{\mathcal{M}}$, and each $\bar{m} \in \bar{\mathcal{M}}$ can be selected multiple times, so that the denominator in (6.62) can be transformed into

$$\prod_{m_i \in M} \sum\nolimits_{\bar{m}_x \in \bar{\mathcal{M}}} e^{-\frac{\varepsilon}{2}d(m_i,\bar{m}_x)} = \sum\nolimits_{\hat{M}_x} e^{-\frac{\varepsilon}{2}d(M,\hat{M}_x)}, \tag{6.63}$$

where $\hat{M}_x$ is the possible sequence of driving statuses that consists of multiple $\hat{m}_x$.

Based on the definition of $d(M_1, M_2)$ in (6.1), we have

$$\prod_{m_i \in M} e^{-\frac{\varepsilon}{2}d(m_i,\hat{m}_i)} = e^{-\frac{\varepsilon}{2}d(M,\hat{M})}. \tag{6.64}$$

Thus, (6.62) can be rewritten as

$$\Pr[\hat{M}|M] = \frac{e^{-\frac{\varepsilon}{2}d(M,\hat{M})}}{\sum_{\hat{M}_x} e^{-\frac{\varepsilon}{2}d(M,\hat{M}_x)}}. \tag{6.65}$$

6.6 Conclusion

In this chapter, we define a unified privacy-preserving measure, i.e., T-I, for mechanisms that use trajectory obfuscation and pseudonym swapping and satisfy ε-DP. Using the T-I, we introduce a new JTOPS, which is the first attempt to jointly employ trajectory obfuscation and pseudonym swapping in road networks to protect both the identity and location privacy of vehicles. The JTOPS has been proven to achieve the T-I and validated with comprehensive experiments. The JTOPS can provide privacy protection in the case of GPAs with full prior knowledge or collusion attacks.

7

Conclusion

In this chapter, we give a conclusion of this book and present the future directions in the related research field.

The robustness of location privacy protection mechanisms in vehicular networks is a critical factor determining whether attackers can compromise the system. Attackers with significant computational resources, extensive knowledge of the system, and access to large datasets are particularly threatening. These sophisticated attackers can potentially break weaker privacy schemes, especially those that rely on limited pseudonym pools or simple anonymization methods.

The design of the system plays a crucial role in defending against such attacks. Advanced cryptographic techniques, frequent changes in pseudonyms, and robust anonymization methods can significantly increase the difficulty for attackers to breach the system. Yet, real-world constraints such as the need for efficient, real-time communication in vehicular networks can introduce vulnerabilities. Systems that prioritize speed over security may reduce privacy protection to maintain performance, providing attackers with potential entry points. Thus, the effectiveness of an attack is a complex interplay between the attacker's capabilities, the strength of the privacy protection mechanisms, and the practical limitations of the system's design.

7.1 Lessons Learned and Open Challenges for LPPM in Future Vehicular Networks

This section starts with illustrating the localization techniques in future vehicular networks, which are compared with their counterparts in current versions. By discussing the new requirements of location privacy and data utility introduced by the novel localization, we present the limitations of the existing LPPMs in future vehicular networks. The potential location privacy issues under the future cross-layer tracking techniques introduced by new communication technologies are illustrated, and we offer solutions to these issues by improving the existing LPPMs.

7.1.1 Advancement of Localization vs. Location Privacy

Localization technologies in future 5G/6G-enabled vehicular networks can be classified as basic localization, cooperative localization, machine learning-based localization, driver tracking, and multipath exploiting localization [185], as shown in Table 7.1. The differences in localization techniques between the future vehicular networks and the current ones are also shown in Table 7.1. We compare the BS density, recourse cost, and localization precision of the localization techniques in Table 7.1. In future vehicular networks, localization, sensing,

DOI: 10.1201/9781003518044-7

and communication will coexist, sharing the same time-frequency-spatial resources [63, 338]. The localization technologies in future vehicular networks are as follows [63].

TABLE 7.1: Difference of localization techniques in current vehicular networks and future vehicular networks.

Localizat-ion Techniques	**Features of Future Vehicular Networks**	**Corresponding Techniques in Current Vehicular Networks**	**Difference**
Traditional localization	High data rates High scalability	Cellular radio-based tracking	More BS and UE More frequent signal transmission More accurate localization
Cooperative localization	Ultra-low latency High scalability	Sensing infrastructure-based tracking Cellular radio-based tracking Upper-layer message-based tracking	More sensors More frequent signal transmission Closer BS and UE
AI-based localization	High data rates High scalability	Sensing infrastructure-based tracking Optical vision-based tracking Upper-layer message-based tracking	More sensors More detailed dataset Massive devices
Channel charting	High scalability	Cellular radio-based tracking Sensing infrastructure-based tracking Upper-layer message-based tracking	Considering channel state information

continued on next page

continued from previous page			
Driver tracking localization	High data rates Ultra-low latency	Sensing infrastructure-based tracking Upper-layer message-based tracking	More sensors 3D vehicular network
Multipath exploiting localization	Ultra-low latency High scalability	Cellular radio-based tracking	More sensors
SLAM	High data rates Ultra-low latency High scalability	Vehicle driving log-based tracking Upper layer message-based tracking Cellular radio-based tracking	More sensors Considering time-varying states More fixed landmarks
Context-aware localization	High data rates Ultra-low latency	Vehicle driving log-based tracking Upper-layer message-based tracking	Intelligent prediction Massive personal information

7.1.1.1 Traditional Localization

Traditional localization technologies mainly focus on geometric calculation with augmented assistance, under the assumption that the positions of the Base Stations (BSs) are knowable and the positions of User Equipment (UEs) are geometric constrained [320]. For example, the position of UEs can be calculated if the distances and physical angular orientations between BSs and UEs are known. Compared with the localization precision of the existing cellular radio-based tracking, the future basic localization could have high precision due to the high data rates and scalability. The frequent communication and the high BS density also expose a granularity trajectory. The traditional localization techniques can be classified into distance, angle, area, and hop-count measurements [193], as follows.

- *Distance Measurement:* Distance measurement calculates the positions with the obtained distance-related measurements. The distance-related measurements leverage transmission time (e.g., time difference of arrival), received signal strength indicator, and connectivity condition [39].
- *Angle Measurement:* Angle measurement employs trigonometry and geometric calculation, which requires the distance and angular information between the target and BSs [423].

• *Area Measurement:* Area measurement localization technology uses the center intersection of all overlapping coverage regions as the estimation position [301]. The more restricted zones are obtained, the higher precision can be achieved [264].

• *Hop-count Measurement:* This kind of localization can be realized by analyzing the number of hops [303, 466].

7.1.1.2 Cooperative Localization

Cooperative localization has been employed in current vehicular networks, as the vehicles and other devices can communicate with others directly through D2D communication protocols [416]. Cooperative localization allows UEs to measure the distance and angular information on each D2D link [291]. UEs are typically closer to each other than to the BSs in D2D communication, which has a higher signal-to-noise ratio [203]. This method provides a high position precision for cooperative localization. Due to the ultra-low latency and high scalability, massive nodes in future vehicular networks will be connected to the network, and the distance between UEs will be closer than that in the current version. Hence, the precision of cooperative localization would be significantly more accurate in future vehicular networks than in the current version.

7.1.1.3 AI-Based Localization

AI-based localization is a data-centric technique [185]. The localization algorithms utilize machine learning to generate a fingerprint database of the environment [13]. The fingerprinting database contains the channel parameters (e.g., channel state information) that are measured at the known locations (reference points) [362]. Then, the localization algorithms can compare the information of the environment surrounding vehicles to estimate the drivers' positions. In future vehicular networks, the dataset for machine learning could be more detailed due to features like hierarchical coexistence, flexible storage, and flexible processing.

7.1.1.4 Channel Charting

Channel charting localization employs AI-based algorithms [407]. It generates a virtual map based on the gained channel state information. The drivers can be located and tracked on the virtual map. Although the map cannot provide the drivers' actual locations, it offers a real-time pseudo-location as reference [63].

TABLE 7.2: Localization technologies in future vehicular networks.

Localization	BS Density	Cost	Precision
Traditional	High	Low [399]	Low [225]
Cooperative	Low	Low [440]	Medium [245]
AI-based	Medium	High [295]	High [295]
Channel Charting	Depends	Depends	Depends
Driver Tracking	Medium	Low [206]	Medium [206]
Multipath Exploiting	Low	Medium [428]	High [201]
SLAM	Low	Depends	High [318]
Context-aware	High	Medium [304]	High [82]

7.1.1.5 Driver Tracking Localization

Driver tracking localization continuously infers the driver's position to smooth out the estimation errors [185]. This localization technique can predict the drivers' trajectories with the vehicle sensor data as follows.

- *Passive Sensing:* Passive sensing in future vehicular networks is also known as passive radar or passive coherent location [63]. Passive sensing locates the target by receiving and processing the energy reflected by the targets [437]. In future sensing infrastructure-based vehicular networks, it is almost impossible to avoid disclosing location privacy by sensor signals.
- *Active Sensing:* Active sensing (radar) localization is employed in various scenarios, e.g., adaptive cruise control and cross-traffic alerts [43]. In future vehicular networks, sensors could share the data related to the location information to support high-precision localization applications. The sensor localization will provide high-precision distance in 3D vehicular networks [63].

7.1.1.6 Multipath-Based Localization

Multipath-based localization discloses the drivers' positions by combining multipath components and environment geometry [427]. The multipath components can be seen as mirror images of a physical environment [164]. Using the road network environment knowledge, multipath-based localization can be launched with the information of the radar signal path. Thus, the multipath-based localization can exploit additional position-related data in radio signals, even if the vehicles are not in the line of sight [124]. As the vehicles in future vehicular networks would be equipped with multiple sensors, the radar-like signals from the sensors could be utilized in multipath-based localization.

7.1.1.7 Simultaneous Localization And Mapping (SLAM)

High data rates, ultra-low latency, and high scalability of the future vehicular networks introduce SLAM for vehicles. SLAM localization aims to estimate the state of the vehicles and the landmarks [83]. Mobile devices and vehicles in SLAM are considered to have time-varying states [318]. In future vehicular networks, the SLAM can be classified into vision-based SLAM and radar-based SLAM as follows.

- *Vision-based SLAM*: It uses image sensors (e.g., cameras) to detect landmarks whose states are fixed or changing slowly.
- *Radar-based SLAM*: It requires laser sensors that can provide higher accuracy than vision-based sensors. 3D lidar point clouds will be widely used in future autonomous vehicular networks.

The entities in future vehicular networks can collect location information with radar-like signals from other vehicles, fixed infrastructures, and sensors [347].

7.1.1.8 Context-Aware Localization

The future 5G/6G-enabled vehicular networks allow intelligent data transmission and multi-modal localization prediction by combining highly personal information with public data [82]. The drivers can select and change communication channels and technologies according to their locations and contexts [304]. Thus, future vehicular networks could utilize the message context in channels for localization.

7.1.2 Limitations and Opportunities of LPPMs for Future Upper Layer Location Privacy Attacks

With the development of vehicular networks, there would be new challenges for the existing LPPMs. The existing LPPMs mainly focus on upper-layer message-based tracking that the adversaries with other tracking techniques (i.e., sensing infrastructure-based, optical vision-based, vehicle driving log-based, and cellular radio-based) cannot be defended in the existing vehicular networks. Nevertheless, as shown in Table 7.3, the existing LPPMs could not provide acceptable location privacy protection under the upper-layer message-based attack scenario in future vehicular networks.

TABLE 7.3: Challenge of location privacy in future vehicular networks.

Techniques	**New Challenges**	**LPPMs That Are Available**	**LPPMs That Need to Be Improved**
THz	Small coverage area Spectrum penetration power Centimeter-level precision via APs	Pass-and-run Certificates for Privacy Secure computation Homomorphic encryption Private information retrieval Searchable encryption Secure communication	Data perturbation Statistical disclosure control Trusted third party
VLC	Signal scatter Observation signal Effective transmission requirement	Certificates for Privacy Secure communication Trusted third party	Pass-and-run Data perturbation Secure computation Statistical disclosure control Private information retrieval Homomorphic encryption Searchable encryption

continued on next page

continued from previous page

mmWave	Exchange CSI frequently Low latency requirement Eavesdrop channel easily	Certificates for Privacy Secure computation Data perturbation Statistical disclosure control Homomorphic encryption Private information retrieval Searchable encryption Trusted third party Secure communication	Pass-and-run
Sub-6 GHz	Flexible antenna design Fake BS and malicious devices In-vehicle tracking	Pass-and-run Certificates for Privacy Secure communication Trusted third party	Secure computation Data perturbation Statistical disclosure control Homomorphic encryption Private information retrieval Searchable encryption
Satellite Communication	Long propagation delay Difficult to allocate MIMO Communicat-ion resource allocation	Certificates for Privacy Data perturbation Statistical disclosure Secure communication Trusted third party	Pass-and-run Secure computation Homomorphic encryption Private information retrieval Searchable encryption

continued on next page

continued from previous page			
QC	Quantum channel Quantum computer	Certificates for Privacy Secure computation Data perturbation Statistical disclosure control Secure communication Trusted third party	Pass-and-run Homomorphic encryption Searchable encryption Private information retrieval

7.1.2.1 User-Side LPPM

Most of the existing user-side LPPMs cannot suit the low latency of the future vehicular networks, i.e., pass-and-run, certificate, and secure computation. Although data perturbation can satisfy the latency requirement, the location data protected by data perturbation cannot provide high-precision LBSs for high-precision localization applications.

• *Pass-and-Run:* The communication delay in the pass-and-run is high since the vehicles apply LBSs by routing through other nodes. Identity verification is also a challenge for pass-and-run-based vehicles because it would lead to high communication delays for legal cooperator authentication [467]. The frequent communication among multiple vehicles could also increase the precision of signal-based localization, i.e., basic localization and cellular radio-based localization.

The transmission delay caused by routing can be decreased by sensors that are fully used in future vehicular networks. With the sensors, the drivers can have full knowledge of the road network and communication environment to select the channels and routers (i.e., vehicles and sensors) [273]. The trust issues of the pass-and-run can be solved by trust management, i.e., authentication and certificate. The advanced blockchain technologies [420] would be utilized in future vehicular networks, which can also be employed for trust management.

• *Certificates for Privacy:* Certificate-based LPPMs achieve authentication by verifying the certifications and signatures [319]. There are two limitations of the certificate-based LPPMs in future vehicular networks: requiring a trusted authority and storage consumption of certificate management [360]. The delay caused by the authentication process might be unacceptable for communication consumption [459]. The massive nodes, i.e., vehicles and sensors, would increase the storage consumption, which would make it challenging to manage certificates.

The features of blockchain, i.e., immutable, secure, and consensus, satisfy the authentication requirement [251]. Therefore, the blockchain platform could be utilized as a support for certificates [58]. The information collected by sensors could also be referred to for authentication.

• *Secure Computation:* The major limitation of secure computation is computation delay, although it satisfies the requirement of flexible storage and processing. Even in current vehicular networks, the computation consumption of the existing secure computation is still a problem [489].

• *Data Perturbation:* Data perturbation is flexible for storage and processing [255]. The communication and calculation delays of the data perturbation are also low. Nevertheless, the future vehicular networks require highly accurate location data, i.e., centimeter-level or millimeter-level, but the existing data-perturbation-based LPPMs generally generate meter-level precision [336]. If the data perturbation provides centimeter-level or millimeter-level precision, the location privacy-preserving capability of the LPPMs could be extremely low [148].

The scenario for data perturbation in future vehicular networks would be limited. Hence, it should be combined with other LPPMs, and adaptive noise should be used to satisfy future scenarios.

7.1.2.2 Server-Side LPPM

The low latency requirement also challenges the existing server-side LPPMs, as the process on the server side could cause a computation delay.

• *Statistical Disclosure Control:* The statistical disclosure control is achieved by using LPPMs, like data perturbation, secure computation, and anonymity on the server side [175]. Therefore, the limitations of the corresponding LPPMs on the user side also exist on the server side. As data transmission in future vehicular networks will be ultra frequent and the amount of data will be huge, computing consumption, storage consumption, and data management will be great challenges for statistical disclosure control.

The data management and storage consumption of statistical disclosure control could be optimized by utilizing blockchain-based techniques, but the computing consumption would be difficult to decrease [373]. AI-based LPPMs can be launched to achieve adaptive protection and efficient analysis to simplify the process [309]. Trust management could also be employed in statistical disclosure control to avoid costing computational resources on trusted entities.

• *Homomorphic Encryption:* Homomorphic encryption is one of the popular, secure computations. The server-side homomorphic encryption is also limited by its computation delay, which can lead to a high computational and communication consumption [144]. It is necessary to simplify the process of homomorphic encryption for computational consumption reduction.

• *Private Information Retrieval:* The computational consumption of the PIR cloud is high, which is difficult to allocate in current vehicular networks. For future vehicular networks with low-latency requirements, the current version of PIR would also be impossible to launch in practice.

• *Searchable Encryption:* The LBSs in future vehicular networks require high-precision location data, but the existing SE schemes would return results with errors to the drivers. The existing works allow the drivers to use SE and other LPPMs together to improve the accuracy of results. Nevertheless, computational consumption can also be increased when SE schemes are combined with certificate-based or other tools.

Information from sensors in future vehicular networks could be referred to correct the result, as massive sensors would be allocated. Also, AI-based LPPMs can be introduced to optimize the result according to the driver's historical information, which would be fine-grained [109].

7.1.2.3 User-Server-Interface LPPM

The user-server-interface LPPMs would attract more researchers' attention in future vehicular networks than in the current version because future vehicular networks would ask the vehicles to share data through frequent V2X communication. Protecting location privacy in channels would be an approach to defend basic localization, channel charting localization, and driver tracking localization.

• *Secure Communication:* The communication channel protocols of future vehicular networks would be improved to be different from the current protocols. As the protocols of current vehicular networks have not been completely developed, we do not consider the protocols in this part. For the end-to-end encrypted services of secure communication, it is seldom used in current vehicular networks [42]. The computational consumption of data encryption and decryption could be high, increasing the latency of communication.

• *Trusted Third Party (TTP):* TTP in current vehicular networks is an ideal scenario, which is difficult to achieve privacy, as the adversaries can hijack the TTP even if the TTP is controlled by government[1]. In future vehicular networks, there will be massive sensors and vehicles that can generate a huge amount of data. A hijacked TTP could lead to serious privacy disclosure. The trust issues of the massive sensors and vehicles also reduce the efficiency of TTP. To solve these limitations, TTP should be combined with other LPPMs, such as encryption, blockchain, AI-based LPPMs, anonymity, and data perturbation.

7.1.3 Location Privacy Challenges and Emerging Wireless Technologies

Different from the existing vehicular networks, the future version transmits data through new communication mediums. The new communication mediums improve the efficiency of vehicular networks but bring new cross-layer location privacy issues.

• *Sub-6 GHz:* The Sub-6 GHz has a spectrum range from 0.45 GHz to 6 GHz and wavelength from 0.15 cm to 2 cm [141]. The wide coverage capability and low cost have been extensively investigated in the existing works to support V2X communications [455].

The sub-6 GHz techniques introduce new challenges to the existing physical-layer LPPMs because of the flexible antenna design [11]. However, the flexible antenna design allows the pass-and-run to transform data effectively. By using fake BSs and malicious devices, the adversaries can obtain distance and angular data by monitoring multiple sub-6 GHz links, with which the adversaries can calculate the positions of vehicles in high precision [1]. The above two issues could be solved by using the certificate and a trusted third party, which aims to detect illegal entities for authentication. Nevertheless, data perturbation and statistical disclosure control increase the difficulty of illegal entity detection.

Malicious applications can be installed on mobile devices or vehicles for tracking the vehicles. The Pegasus developed in Israel can be installed on Android and iOS mobile phones to monitor drivers' location data. The in-vehicle communication via the sub-6 GHz channel can expose the vehicles' location data, as mobile phones have similar location data to the vehicles if the drivers carry mobile phones when driving [64]. To protect location privacy in such a scenario, the driver can hide its semantic information in the location data and block the adversaries' eavesdropping links. Hence, the user-server-interface LPPMs and server-side LPPMs would attract attention to the sub-6 GHz techniques when combined with the AI and blockchain techniques.

• *mmWave:* The mmWave has a spectrum range from 30 GHz to 100 GHz and a wavelength from 1 nm to 10 mm [149]. Due to the beam-based directional transmissions and the utilization of the huge spectrum, mmWave can reach a high multi-gigabit speed and communicate in all weather [120].

In future vehicular networks, the frequent changing of the network topology and channel requires nodes (e.g., vehicles, RSUs, and sensors) to exchange CSI with a low delay, which increases communication consumption [165]. As the V2X communication needs a low

[1]`bleepingcomputer.com/news/security/hacker-claims-to-have-stolen-data-on-1-billion-chinese-citizens/`

transmission delay, the LPPMs in future vehicular networks should be highly efficient. Pass-and-run will increase the transmission delay, so it cannot satisfy the low latency requirement of mmWave.

The adversaries can eavesdrop on the mmWave to obtain and infer the transmitted location data [409]. However, the Doppler shift of signal transmission, which is unavoidable in future mmWave-based vehicular networks, can be combined with the data perturbation and anonymity to protect location privacy [8]. Then, the adversaries cannot obtain the real information by eavesdropping. Secure computation, homomorphic encryption, private information retrieval, and searchable encryption could be used to prevent adversaries from obtaining useful information by eavesdropping because they do not expose the private information in the channel.

• *Terahertz (THz) Communication:* The THz has a spectrum range from 100 GHz to 10 THz and a wavelength from 30 μm to 3000 μm [345]. Nevertheless, the communication region of the THz is small, i.e., less than 50 m [495]. The frameworks of the THz-based vehicular networks have been designed in the existing studies, e.g., [330, 103, 30]. To overcome the limitation of small coverage regions and improve the utilization of THz bands, the design in [330] used an SDN controller to select an optimal route between the source and the destination.

As shown in Table 7.3, the existing LPPMs would face new challenges with THz communication. The pass-and-run mechanism could be improved with the THz-based routing, but the trust issues would be more serious in future vehicular networks than in the current version. Hence, certificate-based LPPMs will be popular in future THz-band communication. Although the low-coverage area and penetration power of the THz spectrum increase the privacy level of the communication medium, THz-band communication requires the vehicles' private information, e.g., whereabouts, driving status, road network information, and traffic condition [330]. The potential optimization for the above two drawbacks could be overcome by combining the trust management mechanisms and blockchain or AI-based techniques. As the coverage area of the THz is small, the data perturbation and statistical disclosure control cannot provide expected location privacy protection with a small number of perturbation candidates. The numerous legal sensors in future networks could also be a good choice for routing the data.

The Access Points (APs) for THz-band communication, which are a kind of trusted third party, can achieve centimeter-level precision of localization when tracking the vehicle's mobility to overcome the limitations of the existing vehicular networks (e.g., frequent network disconnection and volatility of connection) [150]. Nevertheless, the existing DP-based user-side LPPMs cannot satisfy the high-precision localization in such a scenario. Thus, the DP-based LPPMs on the server side have more widespread application prospects than they do on the user side. In future vehicular networks, the user-side DP-based LPPMs should focus on more driving states rather than only using location data to achieve high data utility [253].

• *Visible Light Communication (VLC):* Lighting devices in vehicular networks, such as traffic signals, roadside lights, and vehicle lights, can be employed for VLC to send data with Light-Emitting Diode (LED) installations [367]. VLC has a spectrum range from 400 THz to 800 THz and a wavelength from 380 μm to 780 μm [78]. VLC assists the communication for V2X with low energy consumption and high efficiency [110].

Due to the inevitable scatter issues, VLC outside is not as safe as it is inside [350], and it can be eavesdropped on by adversaries. For outside VLC for V2X, the scattered signals can be received by the adversaries, i.e., eavesdropper [202], even though the techniques like non-orthogonal multiple access [344] and pseudo surface waves [68] can minimize the attenuation effect. Therefore, the pass-and-run in VLC will increase the risk of eavesdropping. Data perturbation and statistical disclosure control cannot protect location data when the source of the signal can be physically observed. As the signal has been observed on the user side, the server-side LPPMs cannot provide location privacy protection.

As the privacy risk introduced by the use of VLC is related to its physical channel characteristics, the certification for authentication and the user-server-interface LPPMs could outperform other existing LPPMs. The certification-based methods can achieve high location privacy protection in short-range VLC by preventing unauthorized receivers from joining VLC, as the scattered signals in shot-range VLC can be ignored [350]. When VLC is allocated for remote communication, the physical layer LPPMs, e.g., cooperative jamming [80] and Channel State Information (CSI) estimation [202], can be used to prevent the eavesdropper from obtaining the transmitted private information. Cooperative jamming is a positive LPPM that aims to block the eavesdropper's channel by allowing multiple legal nodes to route the signal [254]. The CSI estimation is a passive LPPM that asks the legal nodes to infer the CSI of the legal and eavesdropping links with which the legal nodes can select different devices to send signals [202].

• *Quantum Communication (QC):* The QC utilizes the quantum states of lights, which achieves secure communication by using microscopic particles to carry quantum information [474].

Quantum Key Distribution (QKD) has been integrated with existing classical optical networks in QC to achieve cost-efficient and secure communication [356]. Tradition cryptography that relies on mathematical computations is challenged by the development of quantum computing, making QKD important in QC [366]. QKD is based on Heisenberg's uncertainty principle and quantum no-cloning theorem, enabling detection of eavesdropping on the key distribution [354]. However, QKD cannot protect location privacy in vehicular networks, as the laser for QC can be used to infer the direction of the source and destination vehicles. To address this limitation, quantum homomorphic encryption [451], quantum searchable encryption [170], and quantum private information retrieval [24] have been developed for secure QC based on QKD. These techniques enhance the security of QC by enabling computation on encrypted data without decryption, secure search of encrypted data, and retrieval of data from a database without revealing the query contents, respectively.

To enable scalable quantum communication with massive nodes, researchers have been exploring quantum teleportation (QT) in long-range communication scenarios. QT allows for the sender to divide information into traditional and quantum channels and transfer the message to the receiver through both channels, with the message at the sender side being destroyed during transmission. This method has received significant attention due to its potential for log-range communication [305, 306]. However, the use of QT in vehicular networks presents a challenge to existing routing-based PPMs, such as pass-and-run, as they become ineffective in this scenario [306].

Quantum Identity Authentication (QIA) and Quantum Signature (QS) are introduced to realize trust management for secure and reliable QC. The trust management in QC can be classified into objective trust and subjective trust, as follows,

- *Objective Trust:* The trust management is based on certifiable evidence, e.g., certification.
- *Subjective Trust:* The trust management is based on a group with specific characteristics or behaviors, e.g., TTP.

Hence, the certification-based LPPM and statistical disclosure control can be improved to achieve a secure QC in vehicular networks.

• *Reconfigurable Intelligent Surface (RIS):* Comprising passive reflect arrays and control elements, RIS provides a programmable wireless environment for the vehicle-road-human integrated network, which exploits the advantages of the novel electromagnetic wave manipulation technique, i.e., metamaterials [158]. As there are no specific or complicated requirements for the installation location of the metamaterials, RIS can be easily installed, e.g., on the building facades and billboards, which are close to the roads for communication support [18]. The metamaterials would have low computational and energy consumption [135]. Hence, RIS is conducive to future large-scale deployment, which can also conform to the concept of green communication and sustainable development. The support of full-duplex and full-band communication in RIS provides comprehensive coverage and diverse options for vehicular communications [463]. Furthermore, Multi-Input Multi-Output (MIMO) technology could be applied to RIS to utilize the large surface area of RIS for better antenna deployment, which shows great potential in future vehicular networks [47]. However, the mobility and personalization of the drivers in vehicular networks have been overlooked by the existing RIS design [240]. The dynamic network topology and disconnection of vehicular networks lead to a low utility of RIS in vehicular networks, as the RIS cannot gain knowledge from the frequently changed environment or the limited feedback from the drivers [240].

7.1.4 Challenges Arising from Networks Convergence

In the integrated vehicular networks, the vehicles can exchange information and communicate with other vehicles, roadside infrastructure, and pedestrians automatically through real-time V2X communications [117]. With the support of V2X technology, traffic information (e.g., vehicle status, live road conditions, and pedestrian information) enables the formation of the integrated vehicle-road-human network [141]. The components with 5G/6G characteristics in the integrated vehicle-road-human network are illustrated as follows.

• *Integrated Satellite and Terrestrial Network (ISTN):* The satellite communication, which is vulnerable to changeable weather, can provide future vehicular 3D services while cooperating with the existing ground vehicular networks [141].

Satellite communication has limitations such as long propagation delay, difficulty in allocating massive Multiple-Input Multiple-Output (MIMO) networks, and communication resource allocation [455]. Therefore, the pass-and-run is challenging to use in satellite communication because of its transmission delay. Other existing LPPMs, e.g., homomorphic encryption and secure communication, which have high computational consumption and communication delay, should be optimized to improve their efficiency. To overcome the above limitations, the recourse consumption methods could be run on trusted third parties. The data perturbation and statistical disclosure control can protect location privacy effectively so that they can outperform their counterparts in satellite communications.

• *Human Interaction:* The rapid development of mobile devices carried by pedestrians has brought powerful communication and data processing capabilities to enable information interaction between pedestrians and vehicles [69]. Communication between the vehicles and humans (including pedestrians, vehicle drivers and passengers) will be more frequent and efficient in future vehicular networks, which can be utilized to improve the vehicular networks [316]. By sharing data actively and passively with pedestrians, vehicles can overcome the limitations of blind spots in vehicular sensor cameras. Thus, the vehicles can obtain services like accurate situation reporting and early warnings that can reduce traffic accidents and protect lives and property [97]. With mobile devices equipped by drivers and passengers, vehicles can communicate with base stations efficiently and autonomously to complete required tasks. For instance, vehicles without GPS can also be accurately located with the smart devices of passengers [62]. Frequent communication between vehicles and pedestrians increases the risk of cross-layer attacks. The encounter information can be combined to infer the trajectory of the target vehicle. Trust management-based LPPMs can be employed to prevent adversaries who can localize the drivers based on the in-vehicle driving log, and sensing information will be more serious.

7.2 Summary of Outcomes

In this thesis, we started by reviewing the existing localization techniques and LPPMs and discussing their advantages and limitations. We introduced a personalized location privacy-preserving mechanism with a novel Road-Indistinguishability Theorem to balance location privacy protection and data utility in real-world road networks. Based on the RN-I, we then introduced two mechanisms: obfuscated vehicular trajectory detection and cloak region obfuscation. The former allows LBS providers to detect malicious vehicles without breaching the location privacy protection, while the latter extends the parameters of the introduced RN-I. By using anonymity and differential privacy, we introduced a novel Joint Trajectory Obfuscation and pseudonym-swapping mechanism, which is proven to combine two differential privacy processing without introducing the additivity composition theorem of ε-DP.

7.3 Recommendations and Future Research Direction

In the context of an increasingly interconnected digital landscape, the significance of location privacy cannot be overstated. As our reliance on networked systems grows, so too does the need for robust mechanisms to safeguard sensitive location data. Location privacy is particularly critical in an era where ubiquitous computing and the Internet of Things (IoT) seamlessly integrate digital services into our daily lives. The precise tracking capabilities of modern devices pose substantial risks to individual privacy, making the protection of this information a paramount concern.

Existing technologies, while foundational, must be re-evaluated to address the unique characteristics of contemporary networks and the inherent limitations of current methodologies. Traditional approaches to location privacy often rely on static and centralized frameworks that are ill-suited to the dynamic and decentralized nature of future digital ecosystems. These conventional methods are increasingly inadequate in the face of sophisticated adversaries who exploit vulnerabilities in location-sharing services and data aggregation techniques.

A comprehensive approach to enhancing location privacy must integrate several key elements. This includes implementing adaptive privacy policies that can dynamically adjust based on the context of data usage. By considering factors such as user behavior, environmental conditions, and potential threats, these policies can provide more effective and personalized privacy safeguards.

In a nutshell, the future of location privacy depends on our ability to anticipate and respond to the evolving nature of digital threats. By proactively adapting our strategies to the changing landscape, we can better ensure the protection of location privacy in the future. This requires ongoing research, collaboration among stakeholders, and a commitment to developing innovative solutions that prioritize the privacy and security of individuals in the digital age.

Bibliography

[1] Aly Sabri Abdalla, Keith Powell, Vuk Marojevic, and Giovanni Geraci. UAV-assisted attack prevention, detection, and recovery of 5G networks. *IEEE Wirel. Commun.*, 27(4):40–47, 2020.

[2] Haider Rasheed Abdulshaheed, Zeyad Taha Yaseen, Adil M Salman, and Israa Al_Barazanchi. A survey on the use of WiMAX and Wi-Fi on Vehicular Ad-hoc Networks (VANETs). In *IOP Conf. Ser.: Mater. Sci. Eng.*, volume 870, page 012122. IOP Publishing, 2020.

[3] Osman Abul and Cansın Bayrak. From location to location pattern privacy in location-based services. *Knowl. Inf. Sys.*, 56(3):533–557, 2018.

[4] Yemisi Adegoke. Uber drivers in lagos are using a fake gps app to inflate rider fares. *Quartz Africa, November*, 13, 2017.

[5] MH Afifi, Kai Zhou, and Jian Ren. Privacy characterization and quantification in data publishing. *IEEE Trans. Knowl. Data Eng.*, 30(9):1756–1769, 2018.

[6] Ejaz Ahmed and Hamid Gharavi. Cooperative vehicular networking: A survey. *IEEE Trans. Intell. Transp. Syst.*, 19(3):996–1014, 2018.

[7] Nadeem Ahmed, Zhongliang Deng, et al. A survey on location privacy attacks and prevention deployed with IoT in vehicular networks. *Wireless Commun. Mobile Comput.*, 2022, 2022.

[8] Karan Ahuja, Yue Jiang, Mayank Goel, and Chris Harrison. Vid2Doppler: synthesizing doppler radar data from videos for training privacy-preserving activity recognition. In *Proc. CHI Conf. Hum. Factors Comput. Syst. 2021*, pages 1–10, 2021.

[9] Ruqayah Al-ani, Bo Zhou, Qi Shi, Thar Baker, and Mohamed Abdlhamed. Adjusted location privacy scheme for VANET safety applications. In *NOMS 2020-2020 IEEE/IFIP Netw. Oper. Manag. Symp.*, pages 1–4. IEEE, 2020.

[10] Hani Al-Balasmeh, Maninder Singh, and Raman Singh. Framework of data privacy preservation and location obfuscation in vehicular cloud networks. *Concurrency Computat.: Pract. Exper.*, 34(5):e6682, 2022.

[11] Samir Salem Al-Bawri, Md Shabiul Islam, Hin Yong Wong, L Lee, and Mohammad Tariqul Islam. Sub-6 GHz 5G multilayer base station antenna for outdoor localization technique. In *Proc. IEEE Conf. Sustain. Util. Dev. Eng. Technol. CSUDET 2019*, pages 257–260. IEEE, 2019.

[12] Raed Al-Dhubhani and Jonathan M Cazalas. An adaptive geo-indistinguishability mechanism for continuous lbs queries. *Wireless Networks*, 24(8):3221–3239, 2018.

[13] Walaa Y Al-Rashdan and Ashraf Tahat. A comparative performance evaluation of machine learning algorithms for fingerprinting based localization in DM-MIMO wireless systems relying on big data techniques. *IEEE Access*, 8:109522–109534, 2020.

[14] Mohammed Abdullah Ghali Al-Sadoon, Rameez Asif, Yasir Ismael Abdulraheem Al-Yasir, Raed A Abd-Alhameed, and Peter S Excell. AoA localization for vehicle-tracking systems using a dual-band sensor array. *IEEE Trans. Antennas Propag.*, 68(8):6330–6345, 2020.

[15] Mahmood A Al-shareeda, Mohammed Anbar, Iznan H Hasbullah, and Selvakumar Manickam. Survey of authentication and privacy schemes in vehicular ad hoc networks. *IEEE Sens. J.*, 21(2):2422–2433, 2021.

[16] Mahmood A Al-Shareeda, Mohammed Anbar, Selvakumar Manickam, and Iznan H Hasbullah. A secure pseudonym-based conditional privacy-preservation authentication scheme in vehicular ad hoc networks. *Sensors*, 22(5):1696, 2022.

[17] Mahmood A Al-Shareeda, Mohammed Anbar, Selvakumar Manickam, and Ali A Yassin. VPPCS: VANET-based privacy-preserving communication scheme. *IEEE Access*, 8:150914–150928, 2020.

[18] KM Alaaudeen, T Aruna, and G Ananthi. Low-cost, broadband electrically small microstrip antenna using i-shaped metamaterial structure for WLAN/WiMAX and 5G vehicular applications. *Physica B Condens.*, 643:414144, 2022.

[19] Turki Alanzi. A review of mobile applications available in the app and google play stores used during the COVID-19 outbreak. *J. Multidiscip. Healthc.*, 14:45, 2021.

[20] Saad Alharthi, Princy Johnson, and Martin Randles. Secure and energy-efficient communication in IoT/CPS. *Recent Trends in Commun. Netw.*, 2020.

[21] Khatai Aliyev. An explorative analysis of Azerbaijan's COVID-19 policy response and public opinion. *Cauc. Surv.*, 9(3):300–319, 2021.

[22] Yusra Alkendi, Lakmal Seneviratne, and Yahya Zweiri. State of the art in vision-based localization techniques for autonomous navigation systems. *IEEE Access*, 9:76847–76874, 2021.

[23] Tejasvi Alladi, Vinay Chamola, Nishad Sahu, Vishnu Venkatesh, Adit Goyal, and Mohsen Guizani. A comprehensive survey on the applications of blockchain for securing vehicular networks. *IEEE Communications Surveys & Tutorials*, 24(2):1212–1239, 2022.

[24] Matteo Allaix, Seunghoan Song, Lukas Holzbaur, Tefjol Pllaha, Masahito Hayashi, and Camilla Hollanti. On the capacity of quantum private information retrieval from MDS-coded and colluding servers. *IEEE J. Sel. Areas Commun.*, 40(3):885–898, 2022.

[25] Islam Almalkawi, Jafar Raed, Ayoub Alsarhan, Alaa Abdallah, and Emad Abdallah. A novel and efficient priority-based cross-layer contextual unobservability scheme against global attacks for WMSNs. *Int. J. Online Eng.*, 2022.

[26] Aljawharah Alnasser, Hongjian Sun, and Jing Jiang. QoS-balancing algorithm for optimal relay selection in heterogeneous vehicular networks. *IEEE Trans. Intell. Transp. Syst.*, 23(7):8223–8233, 2022.

[27] Ala Altaweel, Hena Mukkath, and Ibrahim Kamel. Gps spoofing attacks in fanets: A systematic literature review. *IEEE Access*, 2023.

[28] Florent Altché and Arnaud de La Fortelle. An LSTM network for highway trajectory prediction. In *2017 IEEE 20th international conference on intelligent transportation systems (ITSC)*, pages 353–359. IEEE, 2017.

[29] Belal Amro. Protecting privacy in VANETs using mix zones with virtual pseudonym change. *arXiv preprint arXiv:1801.10294*, 2018.

[30] G Ananthi and S Sridevi. Stacking dilated convolutional autoencoder beamforming for THz wave vehicular ad-hoc networks. *Wirel. Pers. Commun.*, 126(4):2985–3000, 2022.

[31] Miguel E Andrés, Nicolás E Bordenabe, Konstantinos Chatzikokolakis, and Catuscia Palamidessi. Geo-indistinguishability: Differential privacy for location-based systems. In *Proc. of the 2013 ACM SIGSAC Conf. Comput. commun. secur.*, pages 901–914, 2013.

[32] Miguel E Andrés, Nicolás E Bordenabe, Konstantinos Chatzikokolakis, and Catuscia Palamidessi. Geo-indistinguishability: Differential privacy for location-based systems. In *Proc. 2013 ACM SIGSAC Conf. Comput. Commun. Secur.*, pages 901–914, 2013.

[33] Qasim Ali Arain, Imran Memon, Zhongliang Deng, Muhammad Hammad Memon, Farman Ali Mangi, and Asma Zubedi. Location monitoring approach: multiple mix-zones with location privacy protection based on traffic flow over road networks. *Multimed. Tools. Appl.*, 77(5):5563–5607, 2018.

[34] Karuna Arava and Sumalatha Lingamgunta. Adaptive k-anonymity approach for privacy preserving in cloud. *Arab. J. Sci. Eng.*, 45(4):2425–2432, 2020.

[35] Claudio A Ardagna, Marco Cremonini, Sabrina De Capitani di Vimercati, and Pierangela Samarati. An obfuscation-based approach for protecting location privacy. *IEEE Trans. Depen. Secu. Comput.*, 8(1):13–27, 2009.

[36] Hassan Artail and Noor Abbani. A pseudonym management system to achieve anonymity in vehicular ad hoc networks. *IEEE Trans. Dependable Secure Comput.*, 13(1):106–119, 2015.

[37] Philip Asuquo, Haitham Cruickshank, Jeremy Morley, Chibueze P Anyigor Ogah, Ao Lei, Waleed Hathal, Shihan Bao, and Zhili Sun. Security and privacy in location-based services for vehicular and mobile communications: an overview, challenges, and countermeasures. *IEEE Internet Things J.*, 5(6):4778–4802, 2018.

[38] Giuseppe Ateniese, Daniel H Chou, Breno de Medeiros, and Gene Tsudik. Sanitizable signatures. In *Proc. Eur. Symp. Res. Comput. Secur. (ESORICS)*, pages 159–177. Springer, 2005.

[39] Nur Aisyah Azmi, Shafiqa Samsul, Yoshihide Yamada, Mohd Fitri Mohd Yakub, Mohd Ismifaizul Mohd Ismail, and Rudzidatul Akmam Dziyauddin. A survey of localization using RSSI and TODA techniques in wireless sensor network: System architecture. In *Proc. 2nd Int. Conf. Telematics Future Gen. Netw. (TAFGEN) 2018*, pages 131–136. IEEE, 2018.

[40] Palak Bagga, Ashok Kumar Das, Mohammad Wazid, Joel JPC Rodrigues, and Youngho Park. Authentication protocols in Internet of Vehicles: Taxonomy, analysis, and challenges. *IEEE Access*, 8:54314–54344, 2020.

[41] Alireza Bahramali, Ramin Soltani, Amir Houmansadr, Dennis Goeckel, and Don Towsley. Practical traffic analysis attacks on secure messaging applications. *arXiv preprint arXiv:2005.00508*, 2020.

[42] Wei Bai, Michael Pearson, Patrick Gage Kelley, and Michelle L Mazurek. Improving non-experts' understanding of end-to-end encryption: An exploratory study. In *Proc. IEEE Eur. Symp. Secur. Priv. Wkshp. (EuroS&PW) 2020*, pages 210–219. IEEE, 2020.

[43] Ce Lakpo Bamy, Franck Moukanda Mbango, Dominic Bernard Onyango Konditi, and Pierre Moukala Mpele. A compact dual-band dolly-shaped antenna with parasitic elements for automotive radar and 5G applications. *Heliyon*, 7(4):e06793, 2021.

[44] Rachid El Bansarkhani and Rafael Misoczki. G-Merkle: A hash-based group signature scheme from standard assumptions. In *Proc. Int. Conf. Post-Quantum Cryptogr.*, pages 441–463. Springer, 2018.

[45] Ting Bao, Lei Xu, Liehuang Zhu, Lihong Wang, and Tielei Li. Successive point-of-interest recommendation with personalized local differential privacy. *IEEE Trans. Vel. Technol.*, 70(10):10477–10488, 2021.

[46] Abdullah M Basahel, Adnan Ahmed Abi Sen, Mohammad Yamin, and Sarah Alqahtani. Bartering method for improving privacy of LBS. *Int. J. Comput. Sci. Netw. Secur.*, 19(2):207–213, 2019.

[47] Ertugrul Basar. Reconfigurable intelligent surface-based index modulation: A new beyond mimo paradigm for 6G. *IEEE Trans Commun.*, 68(5):3187–3196, 2020.

[48] Jason Bay, Joel Kek, Alvin Tan, Chai Sheng Hau, Lai Yongquan, Janice Tan, and Tang Anh Quy. BlueTrace: A privacy-preserving protocol for community-driven contact tracing across borders. *Government Technology Agency-Singapore, Tech. Rep*, 2020.

[49] Mohamed Baza, Mahmoud Nabil, Noureddine Lasla, Kemal Fidan, Mohamed Mahmoud, and Mohamed Abdallah. Blockchain-based firmware update scheme tailored for autonomous vehicles. In *Proc. IEEE Wireless Commun. Netw. Conf. (WCNC) 2019*, pages 1–7. IEEE, 2019.

[50] Leila Benarous, Salim Bitam, and Abdelhamid Mellouk. Cslpps: concerted silence-based location privacy preserving scheme for internet of vehicles. *IEEE Trans. Vel. Technol.*, 70(7):7153–7160, 2021.

[51] Leila Benarous and Benamar Kadri. Obfuscation-based location privacy-preserving scheme in cloud-enabled internet of vehicles. *Peer-to-Peer Net. App.*, 15(1):461–472, 2022.

[52] Leila Benarous, Benamar Kadri, and Saadi Boudjit. Alloyed pseudonym change strategy for location privacy in vanets. In *2020 IEEE 17th Annual Consumer Commun. Net. Conf. (CCNC)*, pages 1–6. IEEE, 2020.

[53] Claudio Bettini. Privacy protection in location-based services: A survey. In *Handbook of Mobile Data Privacy*, pages 73–96. Springer, 2018.

[54] Jitendra Bhatia, Yash Modi, Sudeep Tanwar, and Madhuri Bhavsar. Software defined vehicular networks: A comprehensive review. *Int. J. Commun. Syst.*, 32(12):e4005, 2019.

[55] Behnaz Bostanipour and George Theodorakopoulos. Joint obfuscation of location and its semantic information for privacy protection. *Comput. Secur.*, 107:102310, 2021.

[56] Abdelwahab Boualouache and Samira Moussaoui. TAPCS: Traffic-aware pseudonym changing strategy for VANETs. *Peer-to-Peer Netw. Appl.*, 10(4):1008–1020, 2017.

[57] Abdelwahab Boualouache and Samira Moussaoui. Urban pseudonym changing strategy for location privacy in VANETs. *Int. J. Ad Hoc Ubiquitous Comput.*, 24(1-2):49–64, 2017.

[58] Abdelwahab Boualouache, Hichem Sedjelmaci, and Thomas Engel. Consortium blockchain for cooperative location privacy preservation in 5G-enabled vehicular fog computing. *IEEE Trans. Veh. Technol.*, 70(7):7087–7102, 2021.

[59] Abdelwahab Boualouache, Sidi-Mohammed Senouci, and Samira Moussaoui. Towards an efficient pseudonym management and changing scheme for vehicular ad-hoc networks. In *IEEE Glob. Commun. Conf. (GLOBECOM) 2016*, pages 1–7. IEEE, 2016.

[60] Abdelwahab Boualouache, Sidi-Mohammed Senouci, and Samira Moussaoui. A survey on pseudonym changing strategies for vehicular ad-hoc networks. *IEEE Commun. Surv. Tutor.*, 20(1):770–790, 2017.

[61] Azzedine Boukerche and Noura Aljeri. Design guidelines for topology management in software-defined vehicular networks. *IEEE Netw.*, 35(2):120–126, 2021.

[62] Azzedine Boukerche, Burak Kantarci, and Cem Kaptan. Towards ensuring the reliability and dependability of vehicular crowd-sensing data in GPS-less location tracking. *Pervasive Mob. Comput.*, page 101248, 2020.

[63] Andre Bourdoux, Andre Noll Barreto, Barend van Liempd, Carlos de Lima, Davide Dardari, Didier Belot, Elana-Simona Lohan, Gonzalo Seco-Granados, Hadi Sarieddeen, Henk Wymeersch, et al. 6G white paper on localization and sensing. *arXiv preprint arXiv:2006.01779*, 2020.

[64] Matthew Bradbury, Phillip Taylor, Ugur Ilker Atmaca, Carsten Maple, and Nathan Griffiths. Privacy challenges with protecting live vehicular location context. *IEEE Access*, 8:207465–207484, 2020.

[65] Daoud Burghal, Gautam Phadke, Anu Nair, Rui Wang, Tang Pan, Abdullah Algafis, and Andreas F Molisch. Supervised learning approach for relative vehicle localization using V2V MIMO links. In *Proc. IEEE Int. Conf. Commun. (ICC) 2022*, pages 4528–4534. IEEE, 2022.

[66] Ismail Butun and Mikael Gidlund. Location privacy assured Internet of Things. In *Proc. ICISSP*, volume 19, pages 1–8. Prague, Czech Republic, 2019.

[67] Warren Cabral, Craig Valli, Leslie Sikos, and Samuel Wakeling. Review and analysis of cowrie artefacts and their potential to be used deceptively. In *Proc. 2019 Int. Conf. Comp. Sci. Comp. Intell. (CSCI)*, pages 166–171. IEEE, 2019.

[68] Runcheng Cai, Yabin Jin, Timon Rabczuk, Xiaoying Zhuang, and Bahram Djafari-Rouhani. Propagation and attenuation of rayleigh and pseudo surface waves in viscoelastic metamaterials. *J. Appl. Phys.*, 129(12):124903, 2021.

[69] Juan Carmona, Carlos Guindel, Fernando Garcia, and Arturo de la Escalera. eHMI: Review and guidelines for deployment on autonomous vehicles. *Sensors*, 21(9):2912, 2021.

[70] Tânia Carvalho, Nuno Moniz, Pedro Faria, and Luís Antunes. Survey on privacy-preserving techniques for data publishing. *arXiv preprint arXiv:2201.08120*, 2022.

[71] Carlos Rodríguez Casal. Privacy within in-car systems. *Info.*, 7(6):66–75, 2005.

[72] Luomeng Chao, Celimuge Wu, Tsutomu Yoshinaga, Wugedele Bao, and Yusheng Ji. A brief review of multipath TCP for vehicular networks. *Sensors*, 21(8):2793, 2021.

[73] Nirbhay Kumar Chaubey. Security analysis of Vehicular Ad hoc Networks (VANETs): A comprehensive study. *Int. J. Secur. its Appl.*, 10(5):261–274, 2016.

[74] Bhawna Chaudhary and Karan Singh. A blockchain enabled location-privacy preserving scheme for vehicular ad-hoc networks. *Peer-to-Peer Netw. Appl.*, 14(5):3198–3212, 2021.

[75] Madhvi Chaurasia and Bhanu Pratap Singh. Prevention of DoS and routing attack in OLSR under MANET. In *Proc. Int. Conf. Recent Advancement Comp. Commun.*, pages 287–295. Springer, 2018.

[76] Chao Chen, Daqing Zhang, Pablo Samuel Castro, Nan Li, Lin Sun, and Shijian Li. Real-time detection of anomalous taxi trajectories from GPS traces. In *International Conference on Mobile and Ubiquitous Systems: Computing, Networking, and Services*, pages 63–74. Springer, 2011.

[77] Zhenhua Chen, Jingjing Nie, Zhanli Li, Chunpeng Ge, and Willy Susilo. Geometric searchable encryption without false positive and its applications. *Comput. J.*, 2022.

[78] Nan Chi, Yingjun Zhou, Yiran Wei, and Fangchen Hu. Visible light communication in 6G: Advances, challenges, and prospects. *IEEE Veh. Technol. Mag.*, 15(4):93–102, 2020.

[79] Hyunghoon Cho, Daphne Ippolito, and Yun William Yu. Contact tracing mobile apps for COVID-19: Privacy considerations and related trade-offs. *arXiv preprint arXiv:2003.11511*, 2020.

[80] Sunghwan Cho, Gaojie Chen, and Justin P Coon. Cooperative beamforming and jamming for secure VLC system in the presence of active and passive eavesdroppers. *IEEE Trans. Green Commun. Netw.*, 5(4):1988–1998, 2021.

[81] Benny Chor, Oded Goldreich, Eyal Kushilevitz, and Madhu Sudan. Private information retrieval. In *Proc. 36th Annu. IEEE Found. Comput. Sci.*, pages 41–50. IEEE, 1995.

[82] Arsenia Chorti, André Noll Barreto, Stefan Kopsell, Marco Zoli, Marwa Chafii, Philippe Sehier, Gerhard Fettweis, and H Vincent Poor. Context-aware security for 6G wireless the role of physical layer security. *arXiv preprint arXiv:2101.01536*, 2021.

[83] Xinghe Chu, Zhaoming Lu, David Gesbert, Luhan Wang, and Xiangming Wen. Vehicle localization via cooperative channel mapping. *IEEE Trans. Veh. Technol.*, 2021.

[84] William L Croft, Jörg-Rüdiger Sack, and Wei Shi. Obfuscation of images via differential privacy: from facial images to general images. *Peer-to-Peer Netw. Appl.*, 14:1705–1733, 2021.

[85] Lei Cui, Youyang Qu, Mohammad Reza Nosouhi, Shui Yu, Jian-Wei Niu, and Gang Xie. Improving data utility through game theory in personalized differential privacy. *J. Comput. Sci. Technol.*, 34(2):272–286, 2019.

[86] Qimei Cui, Xingxing Hu, Wei Ni, Xiaofeng Tao, Ping Zhang, Tao Chen, Kwang-Cheng Chen, and Martin Haenggi. Vehicular mobility patterns and their applications to Internet-of-Vehicles: A comprehensive survey. *Sci. China Inf. Sci.*, 65(11):1–42, 2022.

[87] Danielle J Currie, Cindy Q Peng, David M Lyle, Brydie A Jameson, and Michael S Frommer. Stemming the flow: How much can the australian smartphone app help to control COVID-19. *Public Health Res. Pract.*, 30(2):3022009, 2020.

[88] Reza Curtmola, Juan Garay, Seny Kamara, and Rafail Ostrovsky. Searchable symmetric encryption: Improved definitions and efficient constructions. *J. Comput. Secur.*, 19(5):895–934, 2011.

[89] Debashis Das, Sourav Banerjee, Uttam Ghosh, Utpal Biswas, and Ali Kashif Bashir. A decentralized vehicle anti-theft system using blockchain and smart contracts. *Peer-to-Peer Netw. Appl.*, 14(5):2775–2788, 2021.

[90] Xinyang Deng, Tianhan Gao, Nan Guo, Jiayu Qi, and Cong Zhao. PAS: Privacy-preserving authentication scheme based on SDN for VANETs. *Appl. Sci.*, 12(9):4791, 2022.

[91] Xinyang Deng, Tianhan Gao, Nan Guo, Cong Zhao, and Jiayu Qi. PCP: A pseudonym change scheme for location privacy preserving in VANETs. *Entropy*, 24(5):648, 2022.

[92] Mahdi Dibaei, Xi Zheng, et al. Attacks and defences on intelligent connected vehicles: A survey. *Digital Communications and Networks*, 6(4):399–421, 2020.

[93] Mahdi Dibaei, Xi Zheng, Youhua Xia, Xiwei Xu, Alireza Jolfaei, Ali Kashif Bashir, Usman Tariq, Dongjin Yu, and Athanasios V Vasilakos. Investigating the prospect of leveraging blockchain and machine learning to secure vehicular networks: A survey. *IEEE Trans. Intell. Transp. Syst.*, 23(2):683–700, 2022.

[94] Tim Dierks and Eric Rescorla. The Transport Layer Security (TLS) protocol version 1.2. In *Internet Eng. Task Force*. Fremont, CA, USA, RFC 5246, August, 2008.

[95] Zhijun Ding, Xiaolun Li, Changjun Jiang, and Mengchu Zhou. Objectives and state-of-the-art of location-based social network recommender systems. *Acm Comput. Surv. (Csur)*, 51(1):1–28, 2018.

[96] Mayank Dixit, Rajesh Kumar, and Anil Kumar Sagar. VANET: Architectures, research issues, routing protocols, and its applications. In *Proc. Int. Conf. Comput. Commun. Autom. (ICCCA) 2016*, pages 555–561. IEEE, 2016.

[97] Joshua E Domeyer, John D Lee, Heishiro Toyoda, Bruce Mehler, and Bryan Reimer. Driver-pedestrian perceptual models demonstrate coupling: Implications for vehicle automation. *IEEE Trans. Hum. Mach. Syst.*, 2022.

[98] John C Duchi, Michael I Jordan, and Martin J Wainwright. Local privacy and statistical minimax rates. In *Proc. 2013 IEEE 54th Ann. Symp. Foundations Comput. Sci. (FOCS)*, pages 429–438. IEEE, 2013.

[99] Matt Duckham and Lars Kulik. Location privacy and location-aware computing. In *Dynamic & Mobile GIS: Investigating Change in Space and Time*, volume 3, pages 35–51. CRC press Boca Raton, Fla, USA, 2006.

[100] Carson Dunbar, Mingze Gao, and Gang Qu. Pass and run: A privacy preserving delay tolerant network communication protocol for cybervehicles. In *2013 Int. Conf. Connect. Veh. Expo. (ICCVE)*, pages 840–841. IEEE, 2013.

[101] Carson Dunbar and Gang Qu. A DTN routing protocol for vehicle location information protection. In *Proc. - IEEE Mil. Commun. Conf. 2014*, pages 94–100. IEEE, 2014.

[102] C. Dwork. Differential privacy. volume 2006, pages 1–12. ICALP, 2006.

[103] Johannes M Eckhardt, Vitaly Petrov, Dmitri Moltchanov, Yevgeni Koucheryavy, and Thomas Kürner. Channel measurements and modeling for low-terahertz band vehicular communications. *IEEE J. Sel. Areas Commun.*, 39(6):1590–1603, 2021.

[104] David Eckhoff, Christoph Sommer, et al. Strong and affordable location privacy in VANETs: Identity diffusion using time-slots and swapping. In *2010 IEEE Vehicular Networking Conference*, pages 174–181. IEEE, 2010.

[105] Mahmoud Hashem Eiza and Qiang Ni. Driving with sharks: Rethinking connected vehicles with vehicle cybersecurity. *IEEE Veh. Technol. Mag.*, 12(2):45–51, 2017.

[106] Emelie Ekenstedt, Lawrence Ong, Yucheng Liu, Sarah Johnson, Phee Lep Yeoh, and Joerg Kliewer. When differential privacy implies syntactic privacy. *IEEE Trans. Inf. Forens. Secur.*, 17:2110–2124, 2022.

[107] Ahmet M Elbir, Gokhan Gurbilek, Burak Soner, Anastasios K Papazafeiropoulos, Pandelis Kourtessis, and Sinem Coleri. Vehicular networks for combating a worldwide pandemic: Preventing the spread of COVID-19. *Smart Health*, 26:100353, 2022.

[108] Abdalla Eldesoky, Ahmed M Kamel, M Elhabiby, and Hadia Elhennawy. Real time localization solution for land vehicle application using low-cost integrated sensors with GPS. *J. Appl. Res. Technol.*, 18(4):214–228, 2020.

[109] Nada M Elfatih, Mohammad Kamrul Hasan, Zeinab Kamal, Deepa Gupta, Rashid A Saeed, Elmustafa Sayed Ali, and Md Sarwar Hosain. Internet of Vehicle's resource management in 5G networks using AI technologies: Current status and trends. *IET Commun.*, 16(5):400–420, 2022.

[110] Noura H ElShabasy and Samy S Soliman. Analysis of multi-hop relaying in VLC-based vehicular networks. In *Proceedings of the 2022 Int. Telecommun. Conf. (ITC-Egypt)*, pages 1–6. IEEE, 2022.

[111] Karim Emara, Wolfgang Woerndl, and Johann Schlichter. On evaluation of location privacy preserving schemes for VANET safety applications. *Comput. Commun.*, 63:11–23, 2015.

[112] Mohamed K Emara, Daniel J King, Hoang V Nguyen, Samer Abielmona, and Shulabh Gupta. Millimeter-Wave slot array antenna front-end for amplitude-only direction finding. *IEEE Trans. Antennas Propag.*, 68(7):5365–5374, 2020.

[113] Richard Gilles Engoulou, Martine Bellaïche, Samuel Pierre, and Alejandro Quintero. VANET security surveys. *Comput. Commun.*, 44:1–13, 2014.

[114] Úlfar Erlingsson, Vitaly Feldman, Ilya Mironov, Ananth Raghunathan, Kunal Talwar, and Abhradeep Thakurta. Amplification by shuffling: From local to central differential privacy via anonymity. In *Proc. 30th Ann. ACM-SIAM Symp. Discrete Algorithms*, pages 2468–2479. SIAM, 2019.

[115] Maryam Faraji-Biregani and Reza Fotohi. Secure communication between UAVs using a method based on smart agents in unmanned aerial vehicles. *J. Supercomput.*, 77(5):5076–5103, 2021.

[116] Fifi Farouk, Yasmin Alkady, and Rawya Rizk. Efficient privacy-preserving scheme for location based services in VANET system. *IEEE Access*, 8:60101–60116, 2020.

[117] Eslam Farsimadan, Francesco Palmieri, Leila Moradi, Dajana Conte, and Beatrice Paternoster. Vehicle-to-Everything (V2X) communication scenarios for Vehicular Ad-hoc Networking (VANET): An overview. In *Int. Conf. Comput. Sci. its Appl.*, pages 15–30. Springer, 2021.

[118] Sharifah Yaqoub Fayi and Zhengguo Sheng. A survey of security, privacy and trust issues in vehicular computation offloading and their solutions using blockchain. *Open Research Europe*, 3, 2023.

[119] Jie Feng, Dening Zeng, Xiuping Jia, Xiangrong Zhang, Jie Li, Yuping Liang, and Licheng Jiao. Cross-frame keypoint-based and spatial motion information-guided networks for moving vehicle detection and tracking in satellite videos. *ISPRS J. Photogramm. Remote Sens.*, 177:116–130, 2021.

[120] Kai-Ten Feng, Li-Hsiang Shen, Chi-Yu Li, Po-Tsang Huang, Sau-Hsuan Wu, Li-Chun Wang, Yi-Bing Lin, and Mau-Chung Frank Chang. 3D on-demand flying mobile communication for millimeter-Wave heterogeneous networks. *IEEE Netw.*, 34(5):198–204, 2020.

[121] Yimeng Feng, Guoqiang Mao, Bo Cheng, Baoqi Huang, Shangbo Wang, and Junliang Chen. MagSpeed: A novel method of vehicle speed estimation through a single magnetic sensor. In *Proc. IEEE Intell. Transp. Syst. Conf. (ITSC) 2019*, pages 4281–4286. IEEE, 2019.

[122] Yimeng Feng, Guoqiang Mao, Bo Cheng, Changle Li, Yilong Hui, Zhigang Xu, and Junliang Chen. MagMonitor: Vehicle speed estimation and vehicle classification through a magnetic sensor. *IEEE Trans. Intell. Transp. Syst.*, 23(2):1311–1322, 2022.

[123] Mohamed Amine Ferrag and Ahmed Ahmim. ESSPR: An efficient secure routing scheme based on searchable encryption with vehicle proxy re-encryption for vehicular peer-to-peer social network. *Telecommun. Syst.*, 66(3):481–503, 2017.

[124] Louis B Fertig, Michael J Baden, J Clayton Kerce, and David Sobota. Localization and tracking with multipath exploitation radar. In *2012 IEEE Radar Conf.*, pages 1014–1018. IEEE, 2012.

[125] Andreas Fuchs, Dustin Kern, Christoph Krauß, and Maria Zhdanova. Securing electric vehicle charging systems through component binding. In *Proc. Int. Conf. Computer Safety, Reliability, and Security (SAFECOMP)*, pages 387–401. Springer, 2020.

[126] N Ganeshkumar and Sanjay Kumar. OBU (On-Board Unit) wireless devices in VANET(s) for effective communication—A review. *Comput. Methods and Data Eng.*, pages 191–202, 2021.

[127] Chen Gao, Chao Huang, Yue Yu, Huandong Wang, Yong Li, and Depeng Jin. Privacy-preserving cross-domain location recommendation. *Proc. of the ACM on Intera. Mobile Wear. Ubi. Technol.*, 3(1):1–21, 2019.

[128] Shangce Gao, Mengchu Zhou, Yirui Wang, Jiujun Cheng, Hanaki Yachi, and Jiahai Wang. Dendritic neuron model with effective learning algorithms for classification, approximation, and prediction. *IEEE Trans. Neural Netw. Learn. Syst.*, 30(2):601–614, 2018.

[129] Zhigang Gao, Yucai Huang, Leilei Zheng, Huijuan Lu, Bo Wu, and Jianhui Zhang. Protecting location privacy of users based on trajectory obfuscation in mobile crowdsensing. *IEEE Trans. Industr. Inform.*, 18(9):6290–6299, 2022.

[130] Paul M Garrett, Yu-Wen Wang, Joshua P White, Yoshihsa Kashima, Simon Dennis, and Cheng-Ta Yang. High acceptance of COVID-19 tracing technologies in Taiwan: A nationally representative survey analysis. *Int. J. Environ. Res. Publich Health*, 19(6):3323, 2022.

[131] William Gasarch. A survey on private information retrieval. *Bulletin of the EATCS*, 82(72-107):113, 2004.

[132] Yong Ge, Hui Xiong, Chuanren Liu, and Zhi-Hua Zhou. A taxi driving fraud detection system. In *2011 IEEE 11th International Conference on Data Mining*, pages 181–190. IEEE, 2011.

[133] Sarah Geber and Thomas N Friemel. A typology-based approach to tracing-app adoption during the COVID-19 pandemic: The case of the SwissCovid app. *J. Quant. Descr.*, 1:online, 2021.

[134] Soheila Ghane, Alireza Jolfaei, Lars Kulik, Kotagiri Ramamohanarao, and Deepak Puthal. Preserving privacy in the Internet of Connected Vehicles. *IEEE Trans. Intell. Transp. Syst.*, 22(8):5018–5027, 2020.

[135] Pratik Ghate and Jonathan Bredow. Quasi-optical beamforming approach using vertically oriented dielectric wedges. *Prog. Electromagn. Res. M*, 105:67–78, 2021.

[136] Răzvan Andrei Gheorghiu, Valentin Iordache, and Valentin Alexandru Stan. Urban traffic detectors-comparison between inductive loop and magnetic sensors. In *Proc. 13th Int. Conf. Electron. Comput. Artif. Intell. (ECAI) 2021)*, pages 1–4. IEEE, 2021.

[137] Sivakanth Gopi, Yin Tat Lee, and Lukas Wutschitz. Numerical composition of differential privacy. *Advances in Neural Information Processing Systems*, 34:11631–11642, 2021.

[138] Kathryn Gray, Daniel Smolyak, Sarkhan Badirli, and George Mohler. Coupled igmm-gans for deep multimodal anomaly detection in human mobility data. *arXiv preprint arXiv:1809.02728*, 2018.

[139] Marco Gruteser and Dirk Grunwald. Anonymous usage of location-based services through spatial and temporal cloaking. In *Proc. 1st Int. Conf. Mobile Syst. Appl. Serv.*, pages 31–42, 2003.

[140] Santanu Guha, Kurt Plarre, Daniel Lissner, Somnath Mitra, Bhagavathy Krishna, Prabal Dutta, and Santosh Kumar. AutoWitness: locating and tracking stolen property while tolerating GPS and radio outages. *ACM Trans. Sens. Netw.*, 8(4):1–28, 2012.

[141] Hongzhi Guo, Xiaoyi Zhou, Jiajia Liu, and Yanning Zhang. Vehicular intelligence in 6G: Networking, communications, and computing. *Veh. Commun.*, 33:100399, 2022.

[142] R. Gupta and U. P. Rao. Achieving location privacy through CAST in location based services. *J. Commun. Netw.*, 19(3):239–249, 2017.

[143] Rajan Gupta, Manan Bedi, Prashi Goyal, Srishti Wadhera, and Vaishnavi Verma. Analysis of COVID-19 tracking tool in India: Case study of Aarogya Setu mobile application. *Digital Government: Research and Practice*, 1(4):1–8, 2020.

[144] Shreya Gupta and Ginni Arora. Use of homomorphic encryption with GPS in location privacy. In *Proc. 4th Int. Conf. ISCON*, pages 42–45. IEEE, 2019.

[145] Achref Haddaji, Samiha Ayed, and Lamia Chaari Fourati. Artificial intelligence techniques to mitigate cyber-attacks within vehicular networks: Survey. *Computers and Electrical Engineering*, 104:108460, 2022.

[146] Sheraz Haider, Deyun Gao, Rehan Ali, Arif Hussain, and Muhammad Talha Ikram. A privacy conserves pseudonym acquisition scheme in vehicular communication systems. *IEEE Trans. Intell. Transp. Syst.*, 23(9):15536–15545, 2022.

[147] Shimaa A Abdel Hakeem and HyungWon Kim. Multi-zone Authentication and Privacy-preserving Protocol (MAPP) based on the bilinear pairing cryptography for 5G-V2X. *Sensors*, 21(2):665, 2021.

[148] Bin Han, Wei Jiang, Mohammad Asif Habibi, and Hans D Schotten. An abstracted survey on 6G: Drivers, requirements, efforts, and enablers. *arXiv preprint arXiv:2101.01062*, 2021.

[149] Chong Han, Yiqin Wang, Yuanbo Li, Yi Chen, Naveed A Abbasi, Thomas Kürner, and Andreas F Molisch. Terahertz wireless channels: A holistic survey on measurement, modeling, and analysis. *IEEE Commun. Surv.*, 24(3):1670–1707, 2022.

[150] Chong Han, Yongzhi Wu, Zhi Chen, Yi Chen, and Guangjian Wang. THz ISAC: A physical-layer perspective of terahertz integrated sensing and communication. *arXiv preprint arXiv:2209.03145*, 2022.

[151] Jun Han, Emmanuel Owusu, Le T Nguyen, Adrian Perrig, and Joy Zhang. ACComplice: Location inference using accelerometers on smartphones. In *Proc. 4th Int. Conf. Commun. Syst. Networks (COMSNETS) 2012*, pages 1–9. IEEE, 2012.

[152] Shuo Han, Ufuk Topcu, and George J Pappas. An approximately truthful mechanism for electric vehicle charging via joint differential privacy. In *Proc. Am. Control Conf. (ACC) 2015*, pages 2469–2475. IEEE, 2015.

[153] Wenlin Han and Yang Xiao. Privacy preservation for V2G networks in smart grid: A survey. *Comput. Commun.*, 91:17–28, 2016.

[154] Ibrahim Abaker Targio Hashem, Victor Chang, Nor Badrul Anuar, Kayode Adewole, Ibrar Yaqoob, Abdullah Gani, Ejaz Ahmed, and Haruna Chiroma. The role of big data in smart city. *Int. J. Inf. Manage.*, 36(5):748–758, 2016.

[155] Hamssa Hasrouny, Abed Ellatif Samhat, Carole Bassil, and Anis Laouiti. VANET security challenges and solutions: A survey. *Veh. Commun.*, 7:7–20, 2017.

[156] Kaiming He, Xiangyu Zhang, Shaoqing Ren, and Jian Sun. Deep residual learning for image recognition. In *Proc. of the IEEE Conference on Computer Vision and Pattern Recognition (CVPR)*, pages 770–778, 6 2016.

[157] Yun He, Jiamin Zhang, Lisha Shuai, Jingtang Luo, Xiaolong Yang, and Qifu Tyler Sun. A personalized secure publishing mechanism of the sensing location data in crowdsensing location-based services. *IEEE Sensors J.*, 2021.

[158] Zhen-Qing He and Xiaojun Yuan. Cascaded channel estimation for large intelligent metasurface assisted massive mimo. *IEEE Wirel. Commun.*, 9(2):210–214, 2019.

[159] Karsten Heimann, Janis Tiemann, Stefan Böcker, and Christian Wietfeld. Cross-bearing based positioning as a feature of 5G millimeter wave beam alignment. In *Proc. IEEE 91st Vehic. Technol. Conf. (VTC2020-Spring)*, pages 1–5. IEEE, 2020.

[160] Csilla Herendy. How were apps developed during, and for, COVID-19?: An investigation into user needs assessment and testing. In *Proc. 11th IEEE Int. Conf. Cogn. Infocommun. (CogInfoCom) 2020*, pages 000503–000508. IEEE, 2020.

[161] Byeongdo Hong, Sangwook Bae, and Yongdae Kim. GUTI reallocation demystified: Cellular location tracking with changing temporary identifier. In *NDSS*, 2018.

[162] Md Jamal Hossain. Impact of COVID-19 pandemic among health care providers in Bangladesh: A systematic review. *Bangladesh J. Infect. Dis.*, pages S8–S15, 2020.

[163] Sara Hsaini, Hayat Bihri, Salma Azzouzi, and My El Hassan Charaf. Contact-tracing approaches to fight COVID-19 pandemic: Limits and ethical challenges. In *Proc. IEEE 2nd Int. Conf. Electron. Control Optim. Comput. Sci. (ICECOCS) 2020*, pages 1–5. IEEE, 2020.

[164] Bin Hu, Zhenyu Shi, and Yi Wang. Single-sensor based indoor localisation by exploiting multipath propagation. *Electron. Lett.*, 54(3):179–181, 2018.

[165] Pengfei Hu, Yifan Ma, Panneer Selvam Santhalingam, Parth H Pathak, and Xiuzhen Cheng. Milliear: Millimeter-Wave acoustic eavesdropping with unconstrained vocabulary. In *IEEE INFOCOM 2022 - IEEE Conf. Comput. Commun.*, pages 11–20. IEEE, 2022.

[166] Jingyu Hua, Wei Tong, Fengyuan Xu, and Sheng Zhong. A geo-indistinguishable location perturbation mechanism for location-based services supporting frequent queries. *IEEE Trans. Inf. Forens.. Secur.*, 13(5):1155–1168, 2017.

[167] Dijiang Huang, Satyajayant Misra, et al. Pacp: An efficient pseudonymous authentication-based conditional privacy protocol for vanets. *IEEE Trans. Intell. Transp. Syst.*, 12(3):736–746, 2011.

[168] Qianyong Huang, Xianyun Xu, Huifang Chen, and Lei Xie. A vehicle trajectory privacy preservation method based on caching and dummy locations in the Internet of Vehicles. *Sensors*, 22(12):4423, 2022.

[169] Rasheed Hussain, Donghyun Kim, Junggab Son, Jooyoung Lee, Chaker Abdelaziz Kerrache, Abderrahim Benslimane, and Heekuck Oh. Secure and privacy-aware incentives-based witness service in social Internet of Vehicles clouds. *IEEE Internet Things J.*, 5(4):2441–2448, 2018.

[170] SK Hafizul Islam, Nimish Mishra, Souvik Biswas, Bharat Keswani, and Sherali Zeadally. An efficient and forward-secure lattice-based searchable encryption scheme for the big-data era. *Comput. Electr. Eng.*, 96:107533, 2021.

[171] Nuraiza Ismail, Syawal Wahid, and Norhayati Ahmad. Arduino based RFID vehicle tracking for home security. In *J. Phys.: Conf. Ser.*, volume 1529, page 022060. IOP Publishing, 2020.

[172] Faisal Jamil, Omar Cheikhrouhou, Harun Jamil, Anis Koubaa, Abdelouahid Derhab, and Mohamed Amine Ferrag. PetroBlock: A blockchain-based payment mechanism for fueling smart vehicles. *Appl. Sci.*, 11(7):3055, 2021.

[173] Satabdy Jena, Narayana Prasad Padhy, and Josep M Guerrero. Multi-layered coordinated countermeasures for dc microgrid clusters under man in the middle attack. *IEEE Transactions on Industry Applications*, 2023.

[174] Baofeng Ji, Yanan Wang, Kang Song, Chunguo Li, Hong Wen, Varun G Menon, and Shahid Mumtaz. A survey of computational intelligence for 6G: Key technologies, applications and trends. *IEEE Trans. Ind. Inform.*, 2021.

[175] Hongbo Jiang, Jie Li, Ping Zhao, Fanzi Zeng, Zhu Xiao, and Arun Iyengar. Location privacy-preserving mechanisms in location-based services: A comprehensive survey. *ACM Comput. Surv. (CSUR)*, 54(1):1–36, 2021.

[176] Jinfang Jiang, Guangjie Han, Hao Wang, and Mohsen Guizani. A survey on location privacy protection in wireless sensor networks. *J. Netw. Comput. Appl.*, 125:93–114, 2019.

[177] Xiantao Jiang, F Richard Yu, Tian Song, and Victor CM Leung. Resource allocation of video streaming over vehicular networks: A survey, some research issues and challenges. *IEEE Trans. Intell. Transp. Syst.*, 23(7):5955–5975, 2022.

[178] Yili Jiang, Kuan Zhang, Yi Qian, and Liang Zhou. Reinforcement-learning-based query optimization in differentially private iot data publishing. *IEEE Internet Things J.*, 8(14):11163–11176, 2021.

[179] Gyuwon Jung, Hyunsoo Lee, Auk Kim, and Uichin Lee. Too much information: Assessing privacy risks of contact trace data disclosure on people with COVID-19 in South Korea. *Public Health Front.*, 8:305, 2020.

[180] Ali El Kaafarani, Essam Ghadafi, and Dalia Khader. Decentralized traceable attribute-based signatures. In *Proc. Cryptograph. Track RSA Conf. (CT-RSA)*, pages 327–348. Springer, 2014.

[181] Nesrine Kaaniche and Maryline Laurent. Attribute-based signatures for supporting anonymous certification. In *Proc. Eur. Symp. Res. Comput. Secur. (ESORICS)*, pages 279–300. Springer, 2016.

[182] Nesrine Kaaniche and Maryline Laurent. BDUA: Blockchain-based data usage auditing. In *Proc. 11th IEEE Int. Conf. CLOUD*, pages 630–637. IEEE, 2018.

[183] Nesrine Kaaniche, Maryline Laurent, and Sana Belguith. Privacy enhancing technologies for solving the privacy-personalization paradox: Taxonomy and survey. *J. Netw. Comput. Appl.*, 171:102807, 2020.

[184] Jian Kang, Doug Steiert, Dan Lin, and Yanjie Fu. Movewithme: Location privacy preservation for smartphone users. *IEEE Trans. Inf. Forens. Secur.*, 15:711–724, 2019.

[185] Ojas Kanhere and Theodore S Rappaport. Position location for futuristic cellular communications: 5G and beyond. *IEEE Commun. Mag.*, 59(1):70–75, 2021.

[186] Gour Karmakar, Abdullahi Chowdhury, Rajkumar Das, Joarder Kamruzzaman, and Syed Islam. Assessing trust level of a driverless car using deep learning. *IEEE Trans. Intell. Transp. Syst.*, 22(7):4457–4466, 2021.

[187] Anabi Hilary Kelechi, Mohammed H Alsharif, Damilare Abdulbasit Oluwole, Philip Achimugu, Osichinaka Ubadike, Jamel Nebhen, Atayero Aaron-Anthony, and Peerapong Uthansakul. The recent advancement in unmanned aerial vehicle tracking antenna: A review. *Sensors*, 21(16):5662, 2021.

[188] Abdhul Khadhir, Himabindu Maripini, Sukanya Sreedhar, Lelitha Vanajakshi, and Bhargava Rama Chilukuri. A study of delay estimation methods at signalized intersections for mixed traffic condition. *Transp. Dev. Econ.*, 7(1):1–16, 2021.

[189] Alla Abbas Khadir and Seyed Amin Hosseini Seno. SDN-based offloading policy to reduce the delay in fog-vehicular networks. *Peer-to-Peer Netw. Appl.*, 14(3):1261–1275, 2021.

[190] Rahan Rasheed Khan, Athar Hanif, and Qadeer Ahmed. Threat analysis of position, navigation, and timing for highly automated vehicles. In *2023 IEEE/ION Position, Location and Navigation Symposium (PLANS)*, pages 647–659. IEEE, 2023.

[191] Shawal Khan, Ishita Sharma, Mazzamal Aslam, Muhammad Zahid Khan, and Shahzad Khan. Security challenges of location privacy in vanets and state-of-the art solutions: A survey. *Future Internet*, 13(4):96, 2021.

[192] Sultan Daud Khan and Habib Ullah. A survey of advances in vision-based vehicle re-identification. *Comput. Vis. Image Underst.*, 182:50–63, 2019.

[193] Fekher Khelifi, Abbas Bradai, Abderrahim Benslimane, Priyanka Rawat, and Mohamed Atri. A survey of localization systems in Internet of Things. *Mob. Netw. Appl.*, 24(3):761–785, 2019.

[194] Mohammad Khodaei and Panos Papadimitratos. Cooperative location privacy in vehicular networks: why simple mix zones are not enough. *IEEE Internet Things J.*, 8(10):7985–8004, 2020.

[195] Paween Khoenkaw. An implementation of programmable oscillator for inductive-loop vehicle sensor using low-cost microcontroller. In *2021 Joint Int. Conf. Digital Arts Media Technol. ECTI North. Sect. Conf. Electr. Electron. Comput. Telecommun. Eng.*, pages 31–36. IEEE, 2021.

[196] Jong Wook Kim, Kennedy Edemacu, and Beakcheol Jang. Privacy-preserving mechanisms for location privacy in mobile crowdsensing: A survey. *J. Netw. Comput. Appl.*, 200:103315, 2022.

[197] Jong Wook Kim, Kennedy Edemacu, Jong Seon Kim, Yon Dohn Chung, and Beakcheol Jang. A survey of differential privacy-based techniques and their applicability to location-based services. *Comput. Secur.*, 111:102464, 2021.

[198] Georgy Kostadinov and Tatiana Atanasova. Security policies for wireless and network infrastructure. In *Probl. Eng. Cybern. Robot.*, number 71, pages 14–19. Bulgarian Academy of Sciences, 2020.

[199] Roman Kostromin. Survey of software configuration management tools of nodes in heterogeneous distributed computing environment. In *Proc. 2nd Int. Workshop ICCS-DE*, pages 156–165, 2020.

[200] Fragkiskos Koufogiannis, Shuo Han, and George J Pappas. Optimality of the laplace mechanism in differential privacy. *arXiv preprint arXiv:1504.00065*, 2015.

[201] Josef Kulmer. High-accuracy positioning exploiting multipath for reducing the infrastructure. *Ph. D. dissertation, Graz University of Technology*, 3, 2019.

[202] Ambrish Kumar, Saif Al-Kuwari, Dushantha Nalin K Jayakody, and Reem Alkanhel. Security performance analysis of a NOMA-assisted underwater VLC system under imprecise channel estimations. *IEEE Access*, 2022.

[203] W. K. Lai, Y. C. Wang, H. C. Lin, and J. W. Li. Efficient resource allocation and power control for LTE-A D2D communication with pure D2D model. *IEEE Trans. Veh. Technol.*, PP(99):1–1, 2020.

[204] Nidhi Lal, Shishupal Kumar, and Vijay Kumar Chaurasiya. A road monitoring approach with real-time capturing of events for efficient vehicles safety in smart city. *Wirel. Pers. Commun.*, 114(1):657–674, 2020.

[205] Marc Langheinrich. Privacy by design—principles of privacy-aware ubiquitous systems. In *Proc. Int. Conf. Ubiquitous Comput.*, pages 273–291. Springer, 2001.

[206] Christos Laoudias, Adriano Moreira, Sunwoo Kim, Sangwoo Lee, Lauri Wirola, and Carlo Fischione. A survey of enabling technologies for network localization, tracking, and navigation. *IEEE Commun. Surv. Tutor.*, 20(4):3607–3644, 2018.

[207] Anna Leander. Parsing pegasus: An infrastructural approach to the relationship between technology and swiss security politics. *Swiss Political Sci. Rev.*, 27(1):205–213, 2021.

[208] Yann LeCun, Léon Bottou, Yoshua Bengio, and Patrick Haffner. Gradient-based learning applied to document recognition. *Proc. of the IEEE*, 86(11):2278–2324, 1998.

[209] Douglas J Leith and Stephen Farrell. Contact tracing app privacy: What data is shared by Europe's GAEN contact tracing apps. In *IEEE INFOCOM 2021 - IEEE Conf. Comput. Commun.*, pages 1–10. IEEE, 2021.

[210] Efrat Levy, Asaf Shabtai, Bogdan Groza, Pal-Stefan Murvay, and Yuval Elovici. CAN-LOC: Spoofing detection and physical intrusion localization on an in-vehicle CAN bus based on deep features of voltage signals. *arXiv preprint arXiv:2106.07895*, 2021.

[211] Bohan Li, Ruochen Liang, Di Zhu, Weitong Chen, and Qinyong Lin. Blockchain-based trust management model for location privacy preserving in VANET. *IEEE Trans. Intell. Transp. Syst.*, 2020.

[212] Gaolei Li, Qiaolun Zhang, Jianhua Li, Jun Wu, and Peng Zhang. Energy-efficient location privacy preserving in vehicular networks using social intimate fogs. *IEEE Access*, 6:49801–49810, 2018.

[213] Hongtao Li, Xingsi Xue, Zhiying Li, Long Li, and Jinbo Xiong. Location privacy protection scheme for LBS in IoT. *Wirel. Commun. Mob. Comput.*, 2021:1–18, 2021.

[214] Hui Li, Lishuang Pei, Dan Liao, Gang Sun, and Du Xu. Blockchain meets VANET: An architecture for identity and location privacy protection in VANET. *Peer-to-Peer Netw. Appl.*, 12(5):1178–1193, 2019.

[215] Jin Li, Man Ho Au, Willy Susilo, Dongqing Xie, and Kui Ren. Attribute-based signature and its applications. In *Proc. 5th ACM Symp. Info. Comput. Commun. Secur.*, pages 60–69, 2010.

[216] Q. Li, H. Wu, L. Liu, B. Pan, and L. Dong. A group based dynamic mix zone scheme for location privacy preservation in VANETs. In *Proc. 3rd Int. Conf. Secur. Smart Cities, Ind. Control Syst. Commun. (SSIC) 2018*, pages 1–5, 2018.

[217] Q. Li, H. Wu, X. Wu, and L. Dong. Multi-level location privacy protection based on differential privacy strategy in VANETs. In *Proc. IEEE 89th Veh. Technol. Conf. (VTC2019-Spring)*, pages 1–5, 2019.

[218] Qingqing Li, Jorge Peña Queralta, Tuan Nguyen Gia, Zhuo Zou, and Tomi Westerlund. Multi-sensor fusion for navigation and mapping in autonomous vehicles: Accurate localization in urban environments. *Unmanned Systems*, 8(03):229–237, 2020.

[219] Qingyuan Li, Hao Wu, Xiang Wu, and Lan Dong. Multi-level location privacy protection based on differential privacy strategy in VANETs. In *2019 IEEE 89th Veh. Technol. Conf. (VTC2019-Spring)*, pages 1–5. IEEE, 2019.

[220] Veronica QT Li, Liang Ma, and Xun Wu. COVID-19, policy change, and post-pandemic data governance: A case analysis of contact tracing applications in East Asia. *Policy Soc.*, 2022.

[221] Wanxin Li, Hao Guo, Mark Nejad, and Chien-Chung Shen. Privacy-preserving traffic management: A blockchain and zero-knowledge proof inspired approach. *IEEE access*, 8:181733–181743, 2020.

[222] X. Li, Y. Ren, L. T. Yang, N. Zhang, B. Luo, J. Weng, and X. Liu. Perturbation-hidden: Enhancement of vehicular privacy for location-based services in Internet of Vehicles. *IEEE Trans. Netw. Sci. Eng.*, pages 1–1, 2020.

[223] X. Li, H. Zhang, Y. Ren, S. Ma, B. Luo, J. Weng, J. Ma, and X. Huang. PAPU: Pseudonym swap with provable unlinkability based on differential privacy in VANETs. *IEEE Internet Things J.*, 7(12):11789–11802, 2020.

[224] Xinghua Li, Huijuan Zhang, Yanbing Ren, Siqi Ma, Bin Luo, Jian Weng, Jianfeng Ma, and Xiaoming Huang. Papu: Pseudonym swap with provable unlinkability based on differential privacy in vanets. *IEEE Internet Things J.*, 7(12):11789–11802, 2020.

[225] Yin-Ya Li, Guo-Qing Qi, and An-Dong Sheng. Performance metric on the best achievable accuracy for hybrid ToA/AoA target localization. *IEEE Commun. Letter*, 22(7):1474–1477, 2018.

[226] Youhuizi Li, Yuyu Yin, et al. A secure dynamic mix zone pseudonym changing scheme based on traffic context prediction. *IEEE Trans. Intell. Transp. Syst.*, 23(7):9492–9505, 2022.

[227] Zi Li, Qingqi Pei, Ian Markwood, Yao Liu, Miao Pan, and Hongning Li. Location privacy violation via GPS-agnostic smart phone car tracking. *IEEE Trans. Veh. Technol.*, 67(6):5042–5053, 2018.

[228] Ruochen Liang, Bohan Li, and Xinyang Song. Blockchain-based privacy preserving trust management model in VANET. In *Proc. Int. Conf. Adv. Data Mining Appl. (ADMA)*, pages 465–479. Springer, 2020.

[229] Wenmin Lin, Xuyun Zhang, Lianyong Qi, Weimin Li, Shancang Li, Victor S Sheng, and Surya Nepal. Location-aware service recommendations with privacy-preservation in the Internet of Things. *IEEE Trans. Comput. Soc. Syst.*, 8(1):227–235, 2020.

[230] Xiaojie Lin, Baihe Ma, Xu Wang, Ying He, Ren Ping Liu, and Wei Ni. Multi-layer reverse engineering system for vehicular controller area network messages. In *Proc. 2022 IEEE 25th Int. Conf. Comput. Support. Coop. Work Des.*, pages 1185–1190. IEEE, 2022.

[231] Mona Naomi Lintvedt. COVID-19 tracing apps as a legal problem: An investigation of the Norwegian'Smittestopp'App. *Oslo L. Rev.*, 8:69, 2021.

[232] B. Liu, W. Zhou, T. Zhu, L. Gao, and Y. Xiang. Location privacy and its applications: A systematic study. *IEEE Access*, 6:17606–17624, 2018.

[233] Bo Liu, Wanlei Zhou, Tianqing Zhu, Longxiang Gao, Tom H Luan, and Haibo Zhou. Silence is golden: Enhancing privacy of location-based services by content broadcasting and active caching in wireless vehicular networks. *IEEE Trans. Veh. Technol.*, 65(12):9942–9953, 2016.

[234] Bo Liu, Wanlei Zhou, Tianqing Zhu, Longxiang Gao, and Yong Xiang. Location privacy and its applications: A systematic study. *IEEE Access*, 6:17606–17624, 2018.

[235] H. Liu, X. Li, H. Li, J. Ma, and X. Ma. Spatiotemporal correlation-aware dummy-based privacy protection scheme for location-based services. In *IEEE INFOCOM 2017 - IEEE Conf. Comput. Commun.*, pages 1–9, 2017.

[236] Haiqing Liu, Yan Zhang, Shiqiang Zheng, and Yuancheng Li. Electric vehicle power trading mechanism based on blockchain and smart contract in V2G network. *IEEE Access*, 7:160546–160558, 2019.

[237] Jianqing Liu, Chi Zhang, Kaiping Xue, and Yuguang Fang. Privacy preservation in multi-cloud secure data fusion for infectious-disease analysis. *IEEE Trans. Mob. Comput.*, early access, 2022.

[238] Jiechao Liu, Paramsothy Jayakumar, Jeffrey L Stein, and Tulga Ersal. Combined speed and steering control in high-speed autonomous ground vehicles for obstacle avoidance using model predictive control. *IEEE Trans. Veh. Technol.*, 66(10):8746–8763, 2017.

[239] Joseph K Liu, Man Ho Au, Tsz Hon Yuen, Cong Zuo, Jiawei Wang, Amin Sakzad, Xiapu Luo, Li Li, and Kim-Kwang Raymond Choo. Privacy-preserving COVID-19 contact tracing app: A zero-knowledge proof approach. *Cryptology ePrint Archive*, 2020.

[240] Yuanwei Liu, Xiao Liu, Xidong Mu, Tianwei Hou, Jiaqi Xu, Marco Di Renzo, and Naofal Al-Dhahir. Reconfigurable intelligent surfaces: Principles and opportunities. *IEEE Commun. Surv. Tutor.*, 23(3):1546–1577, 2021.

[241] Zhaoman Liu, Lei Wu, Junming Ke, Wenlei Qu, Wei Wang, and Hao Wang. Accountable outsourcing location-based services with privacy preservation. *IEEE Access*, 7:117258–117273, 2019.

[242] Zishan Liu, Zhenyu Liu, Lin Zhang, and Xiaodong Lin. Marp: A distributed mac layer attack resistant pseudonym scheme for vanet. *IEEE Trans. Dependable Secure Comput.*, 17(4):869–882, 2018.

[243] Siti-Farhana Lokman, Abu Talib Othman, and Muhammad-Husaini Abu-Bakar. Intrusion detection system for automotive controller area network (CAN) bus system: a review. *EURASIP. J. Wirel. Commun. Netw.*, 2019:1–17, 2019.

[244] Brigitte Lonc, Farah Haidar, and Denis Filatov. Cooperative ITS security standards: Implementation, assessment and next challenges. In *ITS European Congress*, 2020.

[245] Jian Lu, Xu Chen, Maoxin Luo, and Yanran Zhou. Cooperative localization for multiple AUVs based on the rough estimation of the measurements. *Appl. Soft Comput.*, 91:106197, 2020.

[246] Rongxing Lu, Lan Zhang, Jianbing Ni, and Yuguang Fang. 5G vehicle-to-everything services: Gearing up for security and privacy. *Proc. IEEE*, 108(2):373–389, 2019.

[247] Yue Lu and Leiliang Zhang. Social media Wechat infers the development trend of COVID-19. *J. Infect*, 81(1):e82–e83, 2020.

[248] Zhaojun Lu, Mingze Gao, Zhenglin Liu, Gang Qu, and Carson Dunbar. Pass and run: A privacy preserving delay tolerant network communication protocol for cybervehicles. *IEEE Des. Test*, 36(6):56–62, 2019.

[249] Zhaojun Lu, Wenchao Liu, Qian Wang, Gang Qu, and Zhenglin Liu. A privacy-preserving trust model based on blockchain for VANETs. *IEEE Access*, 6:45655–45664, 2018.

[250] Zhaojun Lu, Gang Qu, and Zhenglin Liu. A survey on recent advances in vehicular network security, trust, and privacy. *IEEE Trans. Intell. Transp. Syst.*, 20(2):760–776, 2018.

[251] Bin Luo, Xinghua Li, Jian Weng, Jingjing Guo, and Jianfeng Ma. Blockchain enabled trust-based location privacy protection scheme in VANET. *IEEE Trans. Veh. Technol.*, 69(2):2034–2048, 2019.

[252] Ming Luo and Yanzibo Zhou. An efficient conditional privacy-preserving authentication protocol based on generalized ring signcryption for vanets. *IEEE Trans. Vel. Technol.*, 71(9):10001–10015, 2022.

[253] Baihe Ma, Xiaojie Lin, Xu Wang, Bin Liu, Ying He, Wei Ni, and Ren Ping Liu. New cloaking region obfuscation for road network-indistinguishability and location privacy. In *Proc. 25th Int. Symp. on Res. in Attacks, Intrusions and Defenses (RAID)*, pages 160–170, 2022.

[254] Baihe Ma, Zhihong Liu, Yong Zeng, and Jianfeng Ma. Cooperative jamming for secrecy of wireless communications. In *Proc. Int. Conf. Netw. Netw. Appl. (NaNA) 2018*, pages 14–21. IEEE, 2018.

[255] Baihe Ma, Xu Wang, Wei Ni, and Ren Ping Liu. Personalized location privacy with road network-indistinguishability. *IEEE Trans. Intell. Transp. Syst.*, 23(11), 2022.

[256] Pengchen Ma, Dan Tao, and Tinyu Wu. A pseudonym based anonymous identity authentication mechanism for mobile crowd sensing. In *Proc. 3rd Int. Conf. Big Data Comput. Commun. (BIGCOM)*, pages 10–14. IEEE, 2017.

[257] Wei-Chiu Ma, Ignacio Tartavull, Ioan Andrei Bârsan, Shenlong Wang, Min Bai, Gellert Mattyus, Namdar Homayounfar, Shrinidhi Kowshika Lakshmikanth, Andrei Pokrovsky, and Raquel Urtasun. Exploiting sparse semantic HD maps for self-driving vehicle localization. In *Proc. IEEE/RSJ Int. Conf. Intell. Robot. Syst. IROS 2019*, pages 5304–5311. IEEE, 2019.

[258] Zainab Hikmat Mahmood and Mahmood Khalel Ibrahem. New fully homomorphic encryption scheme based on multistage partial homomorphic encryption applied in cloud computing. In *Proc. 1st Ann. Int. Conf. AiCIS*, pages 182–186. IEEE, 2018.

[259] Gunther Maier. Openstreetmap, the wikipedia map. *REGION*, 1(1):R3–R10, 12 2014.

[260] Aakif Mairaj and Ahmad Y Javaid. Game theoretic solution for an unmanned aerial vehicle network host under ddos attack. *Computer Networks*, 211:108962, 2022.

[261] Hemanta K Maji, Manoj Prabhakaran, and Mike Rosulek. Attribute-based signatures. In *Proc. Cryptograph. Track RSA Conf. (CT-RSA)*, pages 376–392. Springer, 2011.

[262] Nisha Malik, Priyadarsi Nanda, Arushi Arora, Xiangjian He, and Deepak Puthal. Blockchain based secured identity authentication and expeditious revocation framework for vehicular networks. In *Proc. 17th IEEE Int. Conf. Trust, Secur. Privacy Comput. Commun. / 12th IEEE Int. Conf. Big Data Sci. Eng. (TrustCom/BigDataSE)*, pages 674–679. IEEE, 2018.

[263] Jiali Mao, Tao Wang, Cheqing Jin, and Aoying Zhou. Feature grouping-based outlier detection upon streaming trajectories. *IEEE Trans. Knowl. Data Eng.*, 29(12):2696–2709, 2017.

[264] José Ramiro Martinez-de Dios, Alberto de San Bernabé-Clemente, Arturo Torres-González, and Anibal Ollero. Measurement integration for localization and tracking. In *Cluster-based Localization and Tracking in Ubiquitous Computing Systems*, pages 17–50. Springer, 2017.

[265] Elliot Mbunge. Integrating emerging technologies into COVID-19 contact tracing: Opportunities, challenges and pitfalls. *Diabetes Metab. Syndr.: Clin. Res. Rev.*, 14(6):1631–1636, 2020.

[266] Abdueli Paulo Mdee, Malik Muhammad Saad, Murad Khan, Muhammad Toaha Raza Khan, and Dongkyun Kim. Impacts of location-privacy preserving schemes on vehicular applications. *Veh. Commun.*, page 100499, 2022.

[267] Qian Mei, Hu Xiong, Jinhao Chen, Minghao Yang, Saru Kumari, and Muhammad Khurram Khan. Efficient certificateless aggregate signature with conditional privacy preservation in IoV. *IEEE Syst. J.*, 15(1):245–256, 2020.

[268] Imran Memon, Qasim Ali, Asma Zubedi, and Farman Ali Mangi. DPMM: Dynamic pseudonym-based multiple mix-zones generation for mobile traveler. *Multimed. Tools Appl.*, 76(22):24359–24388, 2017.

[269] Ricardo Mendes, Mariana Cunha, and João P Vilela. Impact of frequency of location reports on the privacy level of geo-indistinguishability. *Proc. Priv. Enh.*, 2020(2):379–396, 2020.

[270] Jin Meng, Xufeng Zhang, Tengfei Cao, and Yong Xie. Lightweight and anonymous mutual authentication protocol for IoT devices with physical unclonable functions. *Secur. Commun. Netw.*, 2022:1–11, 2021.

[271] Tobias Meuser, Oluwasegun Taiwo Ojo, Daniel Bischoff, Antonio Fernández Anta, Ioannis Stavrakakis, and Ralf Steinmetz. Hide me: Enabling location privacy in heterogeneous vehicular networks. In *Int. Conf. Netw. Syst.*, pages 11–27. Springer, 2020.

[272] Elham Mohammadzadeh Mianji, Gabriel-Miro Muntean, and Irina Tal. Trustworthy routing in vanet: a q-learning approach to protect against black hole and gray hole attacks. In *2023 IEEE 97th Vehicular Technology Conference (VTC2023-Spring)*, pages 1–6. IEEE, 2023.

[273] Lili Miao, Shang-Fu Chen, Yu-Ling Hsu, and Kai-Lung Hua. How does C-V2X help autonomous driving to avoid accidents? *Sensors*, 22(2):686, 2022.

[274] Qiucheng Miao, Weipeng Jing, and Houbing Song. Differential privacy–based location privacy enhancing in edge computing. *Concurr. Comput. Pract. Exp.*, 31(8):e4735, 2019.

[275] Nader Moayeri, Jalal Mapar, Stefanie Tompkins, and Kaveh Pahlavan. Emerging opportunities for localization and tracking [Guest Editorial]. *IEEE Wirel. Commun.*, 18(2):8–9, 2011.

[276] Mazin Abed Mohammed, Begonya Garcia-Zapirain, Jan Nedoma, Radek Martinek, Prayag Tiwari, Neeraj Kumar, et al. Fully homomorphic enabled secure task offloading and scheduling system for transport applications. *IEEE Trans. Veh. Technol.*, 71(11):12140–12153, 2022.

[277] Rafael Molina-Masegosa, Miguel Sepulcre, Javier Gozalvez, Friedbert Berens, and Vincent Martinez. Empirical models for the realistic generation of cooperative awareness messages in vehicular networks. *IEEE Trans. Veh. Technol.*, 69(5):5713–5717, 2020.

[278] Andrea Motroni, Alice Buffi, Paolo Nepa, and Bernardo Tellini. Sensor-fusion and tracking method for indoor vehicles with low-density UHF-RFID tags. *IEEE Trans. Instrum. Meas.*, 70:1–14, 2020.

[279] Muhammad Akram Mujahid, Kamalrulnizam Abu Bakar, Tasneem SJ Darwish, and Fatima Tul Zuhra. Cluster-based location service schemes in VANETs: Current state, challenges and future directions. *Telecommun. Syst.*, 76(3):471–489, 2021.

[280] Pravin Mundhe, Vijay Kumar Yadav, Abhishek Singh, Shekhar Verma, and S Venkatesan. Ring signature-based conditional privacy-preserving authentication in VANETs. *Wirel. Pers. Commun.*, 114(1):853–881, 2020.

[281] Pravin Mundhe, Vijay Kumar Yadav, Shekhar Verma, and S Venkatesan. Efficient lattice-based ring signature for message authentication in VANETs. *IEEE Syst. J.*, 14(4):5463–5474, 2020.

[282] Hamzah U Mustakim. 5G vehicular network for smart vehicles in smart city: A review. *J. Comput. Electron. Telecommun.*, 1(1):13–18, 2020.

[283] Gopikrishnan Nair, B Anil Kumar, and Lelitha Vanajakshi. Mapping bus and stream travel time using machine learning approaches. *J. Adv. Transp.*, 2022, 2022.

[284] Venkatamangarao Nampally and Dr M Raghavender Sharma. Information sharing standards in communication for VANET. *Int. J. Sci. Res. Comput. Sci. Appl. Manag. Stu.*, 7(4):2319–1953, 2018.

[285] Sashank Narain, Aanjhan Ranganathan, and Guevara Noubir. Security of gps/ins based on-road location tracking systems. In *2019 IEEE Symposium on Security and Privacy (SP)*, pages 587–601. IEEE, 2019.

[286] Snehal Nayak and Meera Narvekar. Real-time vehicle navigation using modified A* algorithm. In *Proc. Int. Conf. Emerg. Trends Innov. ICT (ICEI)*, pages 116–122. IEEE, 2017.

[287] Sen Nie, Ling Liu, and Yuefeng Du. Free-fall: Hacking Tesla from wireless to CAN bus. In *Briefing, Black Hat USA*, volume 25, pages 1–16, 2017.

[288] Nishant Nikhil and Brendan Tran Morris. Convolutional neural network for trajectory prediction. In *Proc. of the European Conference on Computer Vision (ECCV) Workshops*, 2018.

[289] Ben Niu, Yahong Chen, Zhibo Wang, Fenghua Li, Boyang Wang, and Hui Li. Eclipse: Preserving differential location privacy against long-term observation attacks. *IEEE Trans. Mob. Comput.*, 21(1):125–138, 2020.

[290] Ben Niu, Qinghua Li, Xiaoyan Zhu, Guohong Cao, and Hui Li. Achieving k-anonymity in privacy-aware location-based services. In *IEEE INFOCOM 2014 - IEEE Conf. Comput. Commun.*, pages 754–762. IEEE, 2014.

[291] M. Noura and R. Nordin. A survey on interference management for Device-to-Device (D2D) communication and its challenges in 5G networks. *Netw. Comput. Appl.*, pages 130–150, 2016.

[292] Min-hwan Oh and Garud Iyengar. Sequential anomaly detection using inverse reinforcement learning. In *Proc. of the 25th ACM SIGKDD International Conference on Knowledge Discovery & data mining*, pages 1480–1490, 2019.

[293] Eshed Ohn-Bar, Ashish Tawari, Sujitha Martin, and Mohan M Trivedi. On surveillance for safety critical events: In-vehicle video networks for predictive driver assistance systems. *Comput. Vis. Image Underst.*, 134:130–140, 2015.

[294] Tatsuaki Okamoto and Katsuyuki Takashima. Efficient attribute-based signatures for non-monotone predicates in the standard model. *IEEE Trans. Cloud Comput.*, 2(4):409–421, 2014.

[295] Messaoud Ahmed Ouameur, Manouane Caza-Szoka, and Daniel Massicotte. Machine learning enabled tools and methods for indoor localization using low power wireless network. *IEEE Internet Things J.*, page 100300, 2020.

[296] Simon Oya, Carmela Troncoso, and Fernando Pérez-González. Rethinking location privacy for unknown mobility behaviors. In *2019 IEEE European Symposium on Security and Privacy (EuroS&P)*, pages 416–431. IEEE, 2019.

[297] N Padmapriya and S Subathra. A secure communication and location privacy in WSN using grey wolf optimization. In *J. Phys.: Conf. Ser.*, volume 1717, page 012050. IOP Publishing, 2021.

[298] Mark Page and Thanuka L Wickramarathne. Enhanced situational awareness with signals of opportunity: RSS-based localization and tracking. In *Proc. IEEE Intell. Transp. Syst. Conf. (ITSC) 2019*, pages 3833–3838. IEEE, 2019.

[299] David Pàmies-Estrems, Nesrine Kaaniche, Maryline Laurent, Jordi Castellà-Roca, and Joaquin Garcia-Alfaro. Lifelogging protection scheme for internet-based personal assistants. In *Data Privacy Management, Cryptocurrencies and Blockchain Technology*, pages 431–440. Springer, 2018.

[300] Anshul Pandey and Suneel Yadav. Performance evaluation of amplify-and-forward relaying cooperative vehicular networks under physical layer security. *Trans. Emerg. Telecommun. Technol.*, 29(12):e3534, 2018.

[301] Anup Kumar Paul and Takuro Sato. Localization in wireless sensor networks: A survey on algorithms, measurement techniques, applications and challenges. *J. Sens. Actuator Netw*, 6(4):24, 2017.

[302] Andreas Pfitzmann and Marit Hansen. A terminology for talking about privacy by data minimization: Anonymity, unlinkability, undetectability, unobservability, pseudonymity, and identity management, 2010.

[303] Songyut Phoemphon, Chakchai So-In, and Nutthanon Leelathakul. A hybrid localization model using node segmentation and improved particle swarm optimization with obstacle-awareness for wireless sensor networks. *Expert Syst. Appl.*, 143:113044, 2020.

[304] Md Jalil Piran and Doug Young Suh. Learning-driven wireless communications, towards 6G. In *Proc. Int. Conf. Comput. Electron. Commun. Eng. (ICCECE) 2019*, pages 219–224. IEEE, 2019.

[305] Stefano Pirandola, Jens Eisert, Christian Weedbrook, Akira Furusawa, and Samuel L Braunstein. Advances in quantum teleportation. *Nat. Photon.*, 9(10):641–652, 2015.

[306] Stefano Pirandola and Stefano Mancini. Quantum teleportation with continuous variables: A survey. *Laser Phys.*, 16:1418–1438, 2006.

[307] Moritz Platt and Peter McBurney. Sybil in the haystack: a comprehensive review of blockchain consensus mechanisms in search of strong sybil attack resistance. *Algorithms*, 16(1):34, 2023.

[308] Sathe Pooja. Vehicle tracking system using GPS. *Int. J. Sci. Res.*, 2(9):128–130, 2013.

[309] Jason Posner, Lewis Tseng, Moayad Aloqaily, and Yaser Jararweh. Federated learning in vehicular networks: Opportunities and solutions. *IEEE Netw.*, 35(2):152–159, 2021.

[310] R Prathap Kumar, U Srilakshmi, and K Ganesh Reddy. Routing integrity mechanism to prevent wormhole attacks in vehicular adhoc networks. In *Inventive Systems and Control: Proceedings of ICISC 2023*, pages 53–64. Springer, 2023.

[311] NK Prema. Efficient secure aggregation in VANETs using Fully Homomorphic Encryption (FHE). *Mob. Netw. Appl.*, 24(2):434–442, 2019.

[312] Vincent Primault, Antoine Boutet, Sonia Ben Mokhtar, and Lionel Brunie. The long road to computational location privacy: A survey. *IEEE Commun. Surv. Tutor.*, 21(3):2772–2793, 2018.

[313] Tobias Pulls, Roel Peeters, and Karel Wouters. Distributed privacy-preserving transparency logging. In *Proc. 12th ACM Wkshp. Priv. Electr. Soc. (WPES)*, pages 83–94, 2013.

[314] Sarin Valiyaveettil Pushpakaran, Jayakrishnan M Purushothama, Manoj Mani, Aanandan Chandroth, Mohanan Pezholil, and Vasudevan Kesavath. A metamaterial absorber based high gain directional dipole antenna. *Int. J. Microw. Wirel. Technol.*, 10(4):430–436, 2018.

[315] Apostolos Pyrgelis, Nicolas Kourtellis, Ilias Leontiadis, Joan Serrà, and Claudio Soriente. There goes wally: Anonymously sharing your location gives you away. In *2018 IEEE International Conference on Big Data (Big Data)*, pages 1218–1227. IEEE, 2018.

[316] Weijing Qi, Qian Li, Qingyang Song, Lei Guo, and Abbas Jamalipour. Extensive edge intelligence for future vehicular networks in 6G. *IEEE Wirel. Commun.*, 28(4):128–135, 2021.

[317] Yijun Qian, Lijun Yu, Wenhe Liu, and Alexander G Hauptmann. Electricity: An efficient multi-camera vehicle tracking system for intelligent city. In *Proc. IEEE/CVF Conf. Comput. Vis. Pattern Recognit. Workshops*, pages 588–589, 2020.

[318] Tong Qin, Tongqing Chen, Yilun Chen, and Qing Su. Avp-slam: Semantic visual mapping and localization for autonomous vehicles in the parking lot. In *Proc. IEEE/RSJ Int. Conf. Intell. Robot. Syst. IROS 2020*, pages 5939–5945. IEEE, 2020.

[319] Yue Qin, Jingjing Hao, Guangsheng Han, Junjia Pan, Keqiang Han, and Hongyu Niu. Research on secured communication of intelligent connected vehicle based on digital certificate. In *Intell. Autom. Soft Comput.*, pages 1068–1080. Springer, 2021.

[320] Zengyi Qin, Jinglu Wang, and Yan Lu. Monogrnet: A geometric reasoning network for monocular 3D object localization. In *Proc. AAAI Conf. Artif. Intell*, volume 33, pages 8851–8858, 2019.

[321] Chenxi Qiu, Anna Cinzia Squicciarini, Ce Pang, Ning Wang, and Ben Wu. Location privacy protection in vehicle-based spatial crowdsourcing via geo-indistinguishability. *IEEE Trans. Mob. Comput.*, 21(7):2436–2450, 2022.

[322] Youyang Qu, Shui Yu, Wanlei Zhou, and Yonghong Tian. GAN-driven personalized spatial-temporal private data sharing in cyber-physical social systems. *IEEE Trans. Netw. Sci. Eng.*, 7(4):2576–2586, 2020.

[323] Kashif Naseer Qureshi, Adi Alhudhaif, Adeel Abass Shah, Saqib Majeed, and Gwanggil Jeon. Trust and priority-based drone assisted routing and mobility and service-oriented solution for the Internet of Vehicles networks. *J. Inf. Secur. Appl.*, 59:102864, 2021.

[324] Muhammad Wasim Raad, Mohamed Deriche, and Tarek Sheltami. An IoT-based school bus and vehicle tracking system using RFID technology and mobile data networks. *Arab. J. Sci. Eng.*, 46(4):3087–3097, 2021.

[325] Eman Mohammed Radi, Noureddine Lasla, Spiridon Bakiras, and Mohamed Mahmoud. Privacy-preserving electric vehicle charging for peer-to-peer energy trading ecosystems. In *Proc. 2019 IEEE Int. Conf. Commun*, pages 1–6. IEEE, 2019.

[326] Nandana Rajatheva, Italo Atzeni, Emil Bjornson, Andre Bourdoux, Stefano Buzzi, Jean-Baptiste Dore, Serhat Erkucuk, Manuel Fuentes, Ke Guan, Yuzhou Hu, et al. White paper on broadband connectivity in 6G. *arXiv preprint arXiv:2004.14247*, 2020.

[327] Juha Rannikko, Pekka Tamminen, Roosa Hellsten, J Pekka Nuorti, and Jaana Syrjänen. Effectiveness of COVID-19 digital proximity tracing app in Finland. *Clin Microbiol Infect*, 2022.

[328] P Muralidhara Rao and BD Deebak. Security and privacy issues in smart cities/industries: technologies, applications, and challenges. *J. Ambient Intell. Humaniz. Comput.*, pages 1–37, 2022.

[329] Amar Rasheed, Mohamed Baza, Mahmoud Badr, Hani Alshahrani, and Kim-Kwang Raymond Choo. Efficient crypto engine for authenticated encryption, data traceability, and replay attack detection over can bus network. *IEEE Transactions on Network Science and Engineering*, 2023.

[330] Iftikhar Rasheed and Fei Hu. Intelligent super-fast vehicle-to-everything 5G communications with predictive switching between mmWave and THz links. *Veh. Commun.*, 27:100303, 2021.

[331] Jens Helge Reelfs, Oliver Hohlfeld, and Ingmar Poese. Corona-Warn-App: Tracing the start of the official COVID-19 exposure notification app for Germany. In *Proceedings of the SIGCOMM'20 Poster and Demo Sessions*, pages 24–26. 2020.

[332] Eric Rescorla. The Transport Layer Security (TLS) protocol version 1.3. In *Internet Eng. Task Force*. RFC 8446, August, 2018.

[333] Ronald L Rivest, Jon Callas, Ran Canetti, Kevin Esvelt, Daniel Kahn Gillmor, Yael Tauman Kalai, Anna Lysyanskaya, Adam Norige, Ramesh Raskar, Adi Shamir, et al. The PACT protocol specification. *Private Automated Contact Tracing Team, MIT, Cambridge, MA, USA, Tech. Rep. 0.1*, 2020.

[334] Pablo Rodríguez, Santiago Graña, Eva Elisa Alvarez-León, Manuela Battaglini, Francisco Javier Darias, Miguel A Hernán, Raquel López, Paloma Llaneza, Maria Cristina Martín, Oriana Ramirez-Rubio, et al. A population-based controlled experiment assessing the epidemiological impact of digital contact tracing. *Nat. Commun.*, 12(1):1–6, 2021.

[335] Christian Röger, Maja Kalinic, and Jukka M Krisp. Extracting densely covered areas within floating car datasets using inductive loop detector data. *KN-J. Cartogr. Geogr. Inf.*, 71(2):97–103, 2021.

[336] Stefan Roth, Stefano Tomasin, Marco Maso, and Aydin Sezgin. Localization attack by precoder feedback overhearing in 5G networks and countermeasures. *IEEE Trans. Wirel. Commun.*, 20(7):4100–4112, 2021.

[337] Ishtiaq Rouf, Rob Miller, Hossen Mustafa, Travis Taylor, Sangho Oh, Wenyuan Xu, Marco Gruteser, Wade Trappe, and Ivan Seskar. Security and privacy vulnerabilities of in-car wireless networks: A tire pressure monitoring system case study. In *Proc. 19th USENIX Secur. Symp. (USENIX Security 10)*, 2010.

[338] Walid Saad, Mehdi Bennis, and Mingzhe Chen. A vision of 6G wireless systems: Applications, trends, technologies, and open research problems. *IEEE Netw.*, 34(3):134–142, 2019.

[339] Vidyasagar Sadhu, Saman Zonouz, Vincent Sritapan, and Dario Pompili. CollabLoc: Privacy-preserving multi-modal collaborative mobile phone localization. *IEEE Trans. Mob. Comput.*, 20(1):104–116, 2019.

[340] Ikjot Saini, Benjamin St Amour, and Arunita Jaekel. Intelligent adversary placements for privacy evaluation in VANET. *Information*, 11(9):443, 2020.

[341] Krishna Sampigethaya, Mingyan Li, Leping Huang, and Radha Poovendran. Amoeba: Robust location privacy scheme for vanet. *IEEE Journal on Selected Areas in communications*, 25(8):1569–1589, 2007.

[342] Omaji Samuel, Nadeem Javaid, Faisal Shehzad, Muhammad Sohaib Iftikhar, Muhammad Zohaib Iftikhar, Hassan Farooq, and Muhammad Ramzan. Electric vehicles privacy preserving using blockchain in smart community. In *Proc. Int. Conf. Broadband Wireless Comput. Commun. Appl.*, pages 67–80. Springer, 2019.

[343] Abdul Rahman Sani, Muneeb Ul Hassan, and Jinjun Chen. Privacy preserving machine learning for electric vehicles: A survey. *arXiv preprint arXiv:2205.08462*, 2022.

[344] Mubashar Sarfraz, Muhammad Farhan Sohail, Sheraz Alam, Muhammad Javvad ur Rehman, Sajjad Ahmed Ghauri, Khaled Rabie, Hasan Abbas, and Shuja Ansari. Capacity optimization of next-generation UAV communication involving non-orthogonal multiple access. *Drones*, 6(9):234, 2022.

[345] Hadi Sarieddeen, Nasir Saeed, Tareq Y Al-Naffouri, and Mohamed-Slim Alouini. Next generation terahertz communications: A rendezvous of sensing, imaging, and localization. *IEEE Commun. Mag.*, 58(5):69–75, 2020.

[346] Ankur Sarker, Chenxi Qiu, Haiying Shen, Hua Uehara, and Kevin Zheng. Brake data-based location tracking in usage-based automotive insurance programs. In *2020 19th ACM/IEEE Int. Conf. Inf. Proc. Sens. Netw. (IPSN)*, pages 229–240. IEEE, 2020.

[347] Girmi Schouten, Wouter Jansen, and Jan Steckel. Simulation of pulse-echo radar for vehicle control and SLAM. *Sensors*, 21(2):523, 2021.

[348] Émilien Schultz, Rajae Touzani, Julien Mancini, and Jeremy K Ward. From contact tracing to COVID-19 pass holder; the tortured journey of the French TousAntiCovid contact tracing app. *Public Health*, 206:5–7, 2022.

[349] Ramprasaath R. Selvaraju, Michael Cogswell, Abhishek Das, Ramakrishna Vedantam, Devi Parikh, and Dhruv Batra. Grad-cam: Visual explanations from deep networks via gradient-based localization. In *Proc. of the IEEE International Conference on Computer Vision (ICCV)*, pages 618–626, 10 2017.

[350] Rana Shaaban. *Visible Light Communication Cyber Security Vulnerabilities For Indoor And Outdoor Vehicle-To-Vehicle Communication*. PhD thesis, The University of North Dakota, 2021.

[351] Carmel Shachar, Jaclyn Engel, and Glyn Elwyn. Implications for telehealth in a postpandemic future: Regulatory and privacy issues. *Jama*, 323(23):2375–2376, 2020.

[352] Nilay D Shah, Ewout W Steyerberg, and David M Kent. Big data and predictive analytics: Recalibrating expectations. *Jama*, 320(1):27–28, 2018.

[353] Siamak F Shahandashti and Reihaneh Safavi-Naini. Threshold attribute-based signatures and their application to anonymous credential systems. In *Proc. AFRICACRYPT 2009, 2nd Int. Conf. Cryptol. Africa*, pages 198–216. Springer, 2009.

[354] Mehrdad S Sharbaf. Quantum cryptography: a new generation of information technology security system. In *Proc. 6th Int. Conf. Inf. Technol.: New Generations 2009*, pages 1644–1648. IEEE, 2009.

[355] Laxmi Sharma, Abhishek Javali, Rahul Nyamangoudar, R Priya, Pallavi Mishra, and Sudhir K Routray. An update on location based services: Current state and future prospects. In *2017 International Conference on Computing Methodologies and Communication (ICCMC)*, pages 220–224. IEEE, 2017.

[356] Purva Sharma, Anuj Agrawal, Vimal Bhatia, Shashi Prakash, and Amit Kumar Mishra. Quantum key distribution secured optical networks: A survey. *IEEE open j. Commun. Soc.*, 2:2049–2083, 2021.

[357] Surbhi Sharma and Baijnath Kaushik. Applications and challenges in Internet of Vehicles: A survey. *Internet Things its Appl.*, pages 55–65, 2022.

[358] Tanusree Sharma and Masooda Bashir. Use of apps in the COVID-19 response and the loss of privacy protection. *Nat. Med.*, 26(8):1165–1167, 2020.

[359] Tanusree Sharma, Md Mirajul Islam, Anupam Das, SM Taiabul Haque, and Syed Ishtiaque Ahmed. Privacy during pandemic: A global view of privacy practices around COVID-19 apps. In *Proc. ACM SIGCAS Conf. Comput. Sustainable Soc.*, pages 215–229, 2021.

[360] Muhammad Sameer Sheikh, Jun Liang, and Wensong Wang. Security and privacy in vehicular ad hoc network and vehicle cloud computing: a survey. *Wireless Communications and Mobile Computing*, 2020, 2020.

[361] Sheng Shen, Daguan Chen, Yu-Lin Wei, Zhijian Yang, and Romit Roy Choudhury. Voice localization using nearby wall reflections. In *Proc. 26th Ann. Int. Conf. Mobile Comput. Netw.*, pages 1–14, 2020.

[362] Dayu Shi, Xun Zhang, Lina Shi, Andrei Vladimirescu, Wojciech Mazurczyk, Krzysztof Cabaj, Benjamin Meunier, Kareem Ali, John Cosmas, and Yue Zhang. On improving 5G Internet of Radio Light Security based on LED fingerprint identification method. *Sensors*, 21(4):1515, 2021.

[363] Reza Shokri, George Theodorakopoulos, Jean-Yves Le Boudec, and Jean-Pierre Hubaux. Quantifying location privacy. In *2011 IEEE symposium on security and privacy*, pages 247–262. IEEE, 2011.

[364] Navjot Sidhu and Monika Sachdeva. A comprehensive study of routing layer intrusions in zigbee based wireless sensor networks. *International Journal of Advanced Science and Technology*, 29(3):514–524, 2020.

[365] Karen Simonyan and Andrew Zisserman. Very deep convolutional networks for large-scale image recognition. *arXiv preprint arXiv:1409.1556*, 2014.

[366] Amoldeep Singh, Kapal Dev, Harun Siljak, Hem Dutt Joshi, and Maurizio Magarini. Quantum internet—applications, functionalities, enabling technologies, challenges, and research directions. *IEEE Commun. Surv. Tutor.*, 23(4):2218–2247, 2021.

[367] Gurinder Singh, Anand Srivastava, Vivek Ashok Bohara, Md Noor Rahim, Zilong Liu, Dirk Pesch, and Lajos Hanzo. Towards 6G-V2X: Hybrid RF-VLC for vehicular networks. *arXiv preprint arXiv:2208.06287*, 2022.

[368] Pranav Singh, Anup Agarwal, Gaurav Nakum, Danda Rawat, and Sukumar Nandi. MPFSLP: Masqueraded probabilistic flooding for source-location privacy in VANETs. *IEEE Trans. Veh. Technol.*, 69(10):11383–11393, 2020.

[369] Alina Skiljic. 'Stop COVID-19': The croatian application for contact tracing-overview and privacy-related uncertainties. *Eur. Data Prot. L. Rev.*, 6:433, 2020.

[370] Dawn Xiaoding Song, David Wagner, and Adrian Perrig. Practical techniques for searches on encrypted data. In *Proc. 2000 IEEE Symp. Secur. Priv. S&P 2000*, pages 44–55. IEEE, 2000.

[371] Tianyi Song, Nicholas Capurso, et al. Enhancing GPS with lane-level navigation to facilitate highway driving. *IEEE Trans. Vel. Technol.*, 66(6):4579–4591, 2017.

[372] Ridha Soua, Eirini Kalogeiton, Gaetano Manzo, Joao M Duarte, Maria Rita Palattella, Antonio Di Maio, Torsten Braun, Thomas Engel, Leandro A Villas, and Gianluca A Rizzo. SDN coordination for CCN and FC content dissemination in VANETs. In *Ad Hoc Netw.*, pages 221–233. Springer, 2017.

[373] B Sowmiya and E Poovammal. A heuristic k-anonymity based privacy preserving for student management hyperledger fabric blockchain. *Wirel. Pers. Commun.*, 2021:1–18, 2021.

[374] Cynthia Spiess. Is that traffic light tracking you? A case study on a municipal surveillance technology in seattle. *IEEE Trans. Technol. Soc.*, 2(1):15–19, 2021.

[375] Jana Stehlíková et al. The corona crisis, data protection and tracking apps in the EU: The Czech and Austrian COVID-19 mobile phone apps in the battle against the virus. *Mezinárodní vztahy*, 56(1):35–67, 2021.

[376] Michael Stepzinski and Shamik Sengupta. Cybersecurity analysis in dedicated short-range communications in vehicular networks. In *Proc. 11th IEEE Ann. Ubiquitous Comput. Electron. Mobile Commun. Conf. UEMCON*, pages 0021–0027. IEEE, 2020.

[377] Hallam Stevens and Monamie Bhadra Haines. Tracetogether: Pandemic response, democracy, and technology. *Asian Sci. Technol. Soc.*, 14(3):523–532, 2020.

[378] Shijian Su, Houde Dai, Shuying Cheng, Peijie Lin, Chao Hu, and Bowen Lv. A robust magnetic tracking approach based on graph optimization. *IEEE Trans. Instrum. Meas.*, 69(10):7933–7940, 2020.

[379] Shijian Su, Xianping Zeng, Shuang Song, Mingqiang Lin, Houde Dai, Wanan Yang, and Chao Hu. Positioning accuracy improvement of automated guided vehicles based on a novel magnetic tracking approach. *IEEE Intell. Transp. Syst. Mag.*, 12(4):138–148, 2018.

[380] V Subramaniyaswamy, V Jagadeeswari, V Indragandhi, Rutvij H Jhaveri, V Vijayakumar, Ketan Kotecha, and Logesh Ravi. Somewhat homomorphic encryption: Ring learning with error algorithm for faster encryption of IoT sensor signal-based edge devices. *Secur. Commun. Netw.*, 2022, 2022.

[381] SJ Sultanuddin and Md Ali Hussain. Token system-based efficient route optimization in mobile ad hoc network for vehicular ad hoc network in smart city. *Trans. Emerg. Telecommun. Technol.*, 31(12):e3853, 2020.

[382] Gang Sun, Siyu Sun, Hongfang Yu, and Mohsen Guizani. Toward incentivizing fog-based privacy-preserving mobile crowdsensing in the Internet of Vehicles. *IEEE Internet Things J.*, 7(5):4128–4142, 2019.

[383] Hua Sun and Syed Ali Jafar. The capacity of private information retrieval. *IEEE Trans. Inf. Theory.*, 63(7):4075–4088, 2017.

[384] Xiaoqiang Sun, F Richard Yu, Peng Zhang, Weixin Xie, and Xiang Peng. A survey on secure computation based on homomorphic encryption in vehicular ad hoc networks. *Sensors*, 20(15):4253, 2020.

[385] Dajiang Suo, John Moore, Mathew Boesch, Kyle Post, and Sanjay E Sarma. Location-based schemes for mitigating cyber threats on connected and automated vehicles: A survey and design framework. *IEEE Trans. Intell. Transp. Syst.*, 23(4):2919–2937, 2022.

[386] Dajiang Suo, M Elena Renda, and Jinhua Zhao. Quantifying the tradeoff between cybersecurity and location privacy. *arXiv preprint arXiv:2105.01262*, 2021.

[387] Muhammad Naeem Tahir, Marcos Katz, and Urooj Rashid. Analysis of VANET wireless networking technologies in realistic environments. In *Proc. IEEE Radio Wireless Symp.*, pages 123–125. IEEE, 2021.

[388] Shun Takagi, Yang Cao, Yasuhito Asano, and Masatoshi Yoshikawa. Geo-graph-indistinguishability: Protecting location privacy for lbs over road networks. In *IFIP Annual Conf. Data Appl. Secur. Priv.*, pages 143–163. Springer, 2019.

[389] Nazanin Takbiri, Amir Houmansadr, Dennis L Goeckel, and Hossein Pishro-Nik. Matching anonymized and obfuscated time series to users' profiles. *IEEE Trans. Inf. Theory*, 65(2):724–741, 2018.

[390] Hassan Talat and Tuaha andothers Nomani. A survey on location privacy techniques deployed in vehicular networks. In *2019 16th Int. Bhurban Conf. Applied Sci. Technol. (IBCAST)*, pages 604–613. IEEE, 2019.

[391] Zheng Tan, Cheng Wang, Chungang Yan, MengChu Zhou, and Changjun Jiang. Protecting privacy of location-based services in road networks. *IEEE Trans. Intell. Transp. Syst.*, pages 1–14, 2020.

[392] Zheng Tan, Cheng Wang, Mengchu Zhou, and Luomeng Zhang. Private information retrieval in vehicular location-based services. In *Proc. 4th IEEE World Forum on IoT (WF-IoT)*, pages 56–61. IEEE, 2018.

[393] Qi Tang, Jiachen Shen, Zhenfu Cao, and Xiaolei Dong. PSSBP: A privacy-preserving scope-query searchable encryption scheme based on blockchain for parking lots sharing in vehicular networks. In *Proc. 19th IEEE Int. Conf. Embedded Ubiquitous Comput. (EUC)*, pages 1–8. IEEE, 2021.

[394] Haojun Teng, Mianxiong Dong, Yuxin Liu, Wang Tian, and Xuxun Liu. A low-cost physical location discovery scheme for large-scale Internet of Things in smart city through joint use of vehicles and UAVs. *Future Gener. Comput. Syst.*, 118:310–326, 2021.

[395] Sofia Terzi, Charalampos Savvaidis, Konstantinos Votis, Dimitrios Tzovaras, and Ioannis Stamelos. Securing emission data of smart vehicles with blockchain and self-sovereign identities. In *Proc. IEEE Int. Conf. Blockchain 2020*, pages 462–469. IEEE, 2020.

[396] Emmanuel Thomas, MO van Deventer, Thomas Stockhammer, Ali C Begen, M-L Champel, and Ozgur Oyman. Applications and deployments of server and network assisted DASH (SAND). *Proc. Int. Broadcast. Conv. Conf. (IBC)*, page 22, 2016.

[397] Shivendra Tiwari, Saroj Kaushik, Priti Jagwani, and Sunita Tiwari. A survey on LBS: System architecture, trends and broad research areas. In *Proc. Int. Workshop Databases Networked Inf. Syst.*, pages 223–241. Springer, 2011.

[398] Guillermo Tobar, Patricio Galdames, Claudio Gutierrez-Soto, and P Rodriguez-Moreno. A batching location cloaking algorithm for location privacy protection. *Collaborative Technologies and Data Science in Smart City Applications*, pages 26–36, 2018.

[399] Slavisa Tomic, Marko Beko, Rui Dinis, Milan Tuba, and Nebojsa Bacanin. *RSS-AoA-Based Target Localization and Tracking in Wireless Sensor Networks.* River Publishers, 2017.

[400] W. Tong, J. Hua, and S. Zhong. A jointly differentially private scheduling protocol for ridesharing services. *IEEE Trans. Inf. Forensics Secur.*, 12(10):2444–2456, 2017.

[401] Ikram Ullah, Munam A Shah, Abid Khan, and Gwanggil Jeon. Privacy-preserving multilevel obfuscation scheme for vehicular network. *Trans. Emerg. Telecommun. Technol.*, page e4204, 2020.

[402] Ikram Ullah, Munam A Shah, Abid Khan, and Gwanggil Jeon. Privacy-preserving multilevel obfuscation scheme for vehicular network. *Trans. Emerg. Telecommun. Technol.*, 32(2):e4204, 2021.

[403] Ikram Ullah, Munam Ali Shah, Abdul Wahid, Amjad Mehmood, and Houbing Song. ESOT: a new privacy model for preserving location privacy in Internet of Things. *Telecommun. Syst.*, 67(4):553–575, 2018.

[404] Serge Vaudenay. Analysis of DP3T-between Scylla and Charybdis. Technical report, 2020.

[405] Anna Maria Vegni, Carlos Borrego Iglesias, and Valeria Loscri. MOVES: A memory-based vehicular social forwarding technique. *Comput. Netw.*, 197:108324, 2021.

[406] Paula Verde, Javier Díez-González, Rubén Ferrero-Guillén, Alberto Martínez-Gutiérrez, and Hilde Perez. Memetic chains for improving the local wireless sensor networks localization in urban scenarios. *Sensors*, 21(7):2458, 2021.

[407] Harish Viswanathan and Preben E Mogensen. Communications in the 6G era. *IEEE Access*, 8:57063–57074, 2020.

[408] Louis Waked, Mohammad Mannan, and Amr Youssef. To intercept or not to intercept: Analyzing TLS interception in network appliances. In *Proc. 2018 Asia Conf. Comput. Commun. Secur.*, pages 399–412, 2018.

[409] Chao Wang, Feng Lin, Tiantian Liu, Kaidi Zheng, Zhibo Wang, Zhengxiong Li, Ming-Chun Huang, Wenyao Xu, and Kui Ren. mmEve: eavesdropping on smartphone's earpiece via COTS mmWave device. In *Proc. 28th Annu. Int. Conf. Mobile Comput. Netw. (MobiCom)*, pages 338–351, 2022.

[410] Chen Wang, Ping Zhao, Haojun Huang, Rui Zhang, and Weixing Zhu. Privacy-preserving techniques for the 5G-enabled location-based services. In *5G-Enabled Internet of Things*, pages 269–299. CRC Press, 2019.

[411] Danxin Wang, Lan Zhang, Chuanhe Huang, and Xieyang Shen. A privacy-preserving trust management system based on blockchain for vehicular networks. In *Proc. IEEE Wireless Commun. Netw. Conf. (WCNC) 2021*, pages 1–6. IEEE, 2021.

[412] Huajie Wang, Jin Gan, Yong Feng, Yingna Li, and Xiaodong Fu. A privacy enhancement scheme based on blockchain and blind signature for Internet of Vehicles. In *Proc. Int. Conf. Blockchain Trustw. Syst.*, pages 368–387. Springer, 2021.

[413] Jia-Jun Wang and Tufan Kumbasar. Optimal PID control of spatial inverted pendulum with Big Bang–Big Crunch optimization. *IEEE/CAA J. Automat. Sinica*, 7(3):822–832, 2018.

[414] Leye Wang, Dingqi Yang, Xiao Han, Tianben Wang, Daqing Zhang, and Xiaojuan Ma. Location privacy-preserving task allocation for mobile crowdsensing with differential geo-obfuscation. In *Proc. of the 26th Int. Conf. World Wide Web*, pages 627–636, 2017.

[415] Leye Wang, Daqing Zhang, Dingqi Yang, Brian Y Lim, Xiao Han, and Xiaojuan Ma. Sparse mobile crowdsensing with differential and distortion location privacy. *IEEE Trans. Inf. Forens. Secur.*, 15:2735–2749, 2020.

[416] P. Wang, C. M. Chen, S. Kumari, M. Shojafar, and Y. Liu. HDMA: Hybrid D2D Message Authentication scheme for 5G-enabled VANET. *IEEE Trans. Intell. Transp. Syst.*, pages 1–10, 2020.

[417] Shibin Wang and Nianmin Yao. A RSU-aided distributed trust framework for pseudonym-enabled privacy preservation in VANETs. *Wirel. Netw.*, 25(3):1099–1115, 2019.

[418] Shibin Wang, Nianmin Yao, Ning Gong, and Zhenguo Gao. A trigger-based pseudonym exchange scheme for location privacy preserving in VANETs. *Peer-to-Peer Netw. Appl.*, 11(3):548–560, 2018.

[419] Xiaojie Wang, Zhaolong Ning, MengChu Zhou, Xiping Hu, Lei Wang, Yan Zhang, Fei Richard Yu, and Bin Hu. Privacy-preserving content dissemination for vehicular social networks: Challenges and solutions. *IEEE Commun. Surveys Tuts.*, 21(2):1314–1345, 2018.

[420] Xu Wang, Xuan Zha, Wei Ni, Ren Ping Liu, Y Jay Guo, Xinxin Niu, and Kangfeng Zheng. Survey on blockchain for Internet of Things. *Comput. Commun.*, 136:10–29, 2019.

[421] Yang Wang, Sunan Wang, Shengyu Zhang, and Hongjie Cen. An edge-assisted data distribution method for vehicular network services. *IEEE Access*, 7:147713–147720, 2019.

[422] Yingjie Wang, Zhipeng Cai, Xiangrong Tong, Yang Gao, and Guisheng Yin. Truthful incentive mechanism with location privacy-preserving for mobile crowdsourcing systems. *Comput. Netw.*, 135:32–43, 2018.

[423] Yue Wang and KC Ho. Unified near-field and far-field localization for AoA and hybrid AoA-TDOA positionings. *IEEE Trans. Wirel. Commun.*, 17(2):1242–1254, 2017.

[424] Yunling Wang, Jianfeng Wang, and Xiaofeng Chen. Secure searchable encryption: A survey. *J. commun. inf. netw.*, 1(4):52–65, 2016.

[425] Yuhei Watanabe, Hideki Yamamoto, and Hirotaka Yoshida. A study on the applicability of the Lesamnta-LW lightweight hash function to TPMS. *IACR Cryptol. ePrint Arch.*, 2019:840, 2019.

[426] Kang Wei, Jun Li, Ming Ding, Chuan Ma, Howard H Yang, Farhad Farokhi, Shi Jin, Tony QS Quek, and H Vincent Poor. Federated learning with differential privacy: Algorithms and performance analysis. *IEEE Trans. Inf. Forens. Secu.*, 15:3454–3469, 2020.

[427] Fuxi Wen and Henk Wymeersch. 5G synchronization, positioning, and mapping from diffuse multipath. *IEEE Wirel. Commun.*, 10(1):43–47, 2020.

[428] Stijn Wielandt and Lieven De Strycker. Indoor multipath assisted angle of arrival localization. *Sensors*, 17(11):2522, 2017.

[429] Alexander Wood, Kayvan Najarian, and Delaram Kahrobaei. Homomorphic encryption for machine learning in medicine and bioinformatics. *ACM Comput. Surv. (CSUR)*, 53(4):1–35, 2020.

[430] Karel Wouters, Koen Simoens, Danny Lathouwers, and Bart Preneel. Secure and privacy-friendly logging for egovernment services. In *Proc. 3rd Int. Conf. on Availability, Reliability and Security (ARES 2008)*, pages 1091–1096. IEEE, 2008.

[431] Lei Wu, Xia Wei, Lingzhen Meng, Shengnan Zhao, and Hao Wang. Privacy-preserving location-based traffic density monitoring. *Conn Sci.*, 34(1):874–894, 2022.

[432] Libing Wu, Jing Fan, Yong Xie, Jing Wang, and Qin Liu. Efficient location-based conditional privacy-preserving authentication scheme for vehicle ad hoc networks. *Int. J. Distrib. Sens. Netw.*, 13(3):1550147717700899, 2017.

[433] Yonghui Xiao and Li Xiong. Protecting locations with differential privacy under temporal correlations. In *Proc. 22nd ACM SIGSAC Conf. Comput. Commun. Secur.*, pages 1298–1309, 2015.

[434] Yonghui Xiao, Li Xiong, Si Zhang, and Yang Cao. Loclok: Location cloaking with differential privacy via hidden markov model. *Proc. of the VLDB Endowment*, 10(12):1901–1904, 2017.

[435] Zhu Xiao, Pingting Li, Vincent Havyarimana, Georges Maigary Hassana, Dong Wang, and Keqin Li. GOI: A novel design for vehicle positioning and trajectory prediction under urban environments. *IEEE Sens. J.*, 18(13):5586–5594, 2018.

[436] Peng-shou Xie, Xue-ming Han, Tao Feng, Yan Yan, and Guo-qiang Ma. Location privacy protection algorithm based on the division method of voronoi in the Internet of Vehicles. In *Proc. Int. Conf. Artif. Intell. Secur.*, pages 412–422. Springer, 2020.

[437] Rui Xie, Kai Luo, and Tao Jiang. Joint coverage and localization driven receiver placement in distributed passive radar. *IEEE Trans. Geosci. Remote Sens.*, 59(2):1094–1105, 2020.

[438] Jinbo Xiong, Rong Ma, Lei Chen, Youliang Tian, Qi Li, Ximeng Liu, and Zhiqiang Yao. A personalized privacy protection framework for mobile crowdsensing in iiot. *IEEE Trans. Industr. Inform.*, 16(6):4231–4241, 2019.

[439] Zuobin Xiong, Zhipeng Cai, Qilong Han, Arwa Alrawais, and Wei Li. Adgan: protect your location privacy in camera data of auto-driving vehicles. *IEEE Trans. Ind. Inform.*, 17(9):6200–6210, 2020.

[440] Bo XU, Shengxin LI, and Huan ZHANG. Low-cost multi-AUV cooperative localization method based on dual-model. *J. Ship Res.*, 15(3), 2020.

[441] Chugui Xu, Ju Ren, Deyu Zhang, and Yaoxue Zhang. Distilling at the edge: A local differential privacy obfuscation framework for iot data analytics. *IEEE Commun. Mag.*, 56(8):20–25, 2018.

[442] Hao Xu, Lei Zhang, Oluwakayode Onireti, Yang Fang, William J Buchanan, and Muhammad Ali Imran. BeepTrace: Blockchain-enabled privacy-preserving contact tracing for COVID-19 pandemic and beyond. *IEEE Internet Things J.*, 8(5):3915–3929, 2020.

[443] Yehong Xu, Dan He, Pingfu Chao, Jiwon Kim, Wen Hua, and Xiaofang Zhou. Route reconstruction using low-quality bluetooth readings. In *Proc. 28th Int. Conf. Adv. Geogr. Inf. Syst.*, pages 179–182, 2020.

[444] Vijay Kumar Yadav, Nitish Andola, Shekhar Verma, and S Venkatesan. Anonymous and linkable location-based services. *IEEE Trans. Veh. Technol.*, 71(9):9397–9409, 2022.

[445] Vijay Kumar Yadav, Shekhar Verma, and S Venkatesan. Efficient and secure location-based services scheme in VANET. *IEEE Trans. Veh. Technol.*, 69(11):13567–13578, 2020.

[446] Vijay Kumar Yadav, Shekhar Verma, and Subramanian Venkatesan. Linkable privacy-preserving scheme for location-based services. *IEEE Trans. Intell. Transp. Syst.*, 2021.

[447] Hao Yang, Jiarui Cai, Meixin Zhu, Chenxi Liu, and Yinhai Wang. Traffic-Informed Multi-camera Sensing (TIMS) system based on vehicle re-identification. *IEEE Trans. Intell. Transp. Syst.*, 23(10):17189–17200, 2022.

[448] Manxiang Yang, Baopeng Ye, Yuling Chen, Tao Li, Yixian Yang, Xiaobin Qian, and Xiaomei Yu. A trusted de-swinging k-anonymity scheme for location privacy protection. *Journal of Cloud Computing*, 11(1):1–15, 2022.

[449] Andrew C Yao. Protocols for secure computations. In *Proc. 23rd Ann. Symp. Found. Comp. Sci. (SFCS 1982)*, pages 160–164. IEEE, 1982.

[450] Di Yao, Chao Zhang, Zhihua Zhu, Qin Hu, Zheng Wang, Jianhui Huang, and Jingping Bi. Learning deep representation for trajectory clustering. *Expert Syst.*, 35(2):e12252, 2018.

[451] Maxwell Yarter, Glen Uehara, and Andreas Spanias. Implementation and analysis of quantum homomorphic encryption. In *Proc. 13th Int. Conf. Int. Conf. Inf. Intell. Syst. Appl.*, pages 1–5. IEEE, 2022.

[452] Dayong Ye, Sheng Shen, Tianqing Zhu, Bo Liu, and Wanlei Zhou. One parameter defense—defending against data inference attacks via differential privacy. *IEEE Trans. Inf. Forens. Secur.*, 17:1466–1480, 2022.

[453] Qingqing Ye, Haibo Hu, Xiaofeng Meng, and Huadi Zheng. Privkv: Key-value data collection with local differential privacy. In *2019 IEEE Symposium on Security and Privacy (SP)*, pages 317–331. IEEE, 2019.

[454] Tatu Ylönen and Chris Lonvick. The Secure Shell (SSH) protocol architecture, 2006.

[455] Xiaohu You, Cheng-Xiang Wang, Jie Huang, Xiqi Gao, Zaichen Zhang, Mao Wang, Yongming Huang, Chuan Zhang, Yanxiang Jiang, Jiaheng Wang, et al. Towards 6G wireless communication networks: Vision, enabling technologies, and new paradigm shifts. *Sci. China Inf. Sci.*, 64(1):1–74, 2021.

[456] Guangsheng Yu, Xuan Zha, Xu Wang, Wei Ni, Kan Yu, Ping Yu, J Andrew Zhang, Ren Ping Liu, and Y Jay Guo. Enabling attribute revocation for fine-grained access control in blockchain-IoT systems. *IEEE Trans. Eng. Manage.*, 67(4):1213–1230, 2020.

[457] Haining Yu, Hongli Zhang, Xiangzhan Yu, Xiaojiang Du, and Mohsen Guizani. PGRide: Privacy-preserving group ridesharing matching in online ride hailing services. *IEEE Internet Things J.*, 8(7):5722–5735, 2020.

[458] Lei Yu, Ling Liu, and Calton Pu. Dynamic differential ;ocation privacy with personalized error bounds. In *Network and Distributed System Security Symposium (NDSS)*, 2017.

[459] Ping Yu, Wei Ni, Guangsheng Yu, Hua Zhang, Ren Ping Liu, and Qiaoyan Wen. Efficient anonymous data authentication for vehicular ad hoc networks. *Secur. Commun. Netw.*, 2021, 2021.

[460] Rong Yu, Jiawen Kang, et al. Mixgroup: Accumulative pseudonym exchanging for location privacy enhancement in vehicular social networks. *IEEE Trans. Dependable Secure Comput.*, 13(1):93–105, 2015.

[461] Yinghao Yu, Jun Zhang, and Khaled B Letaief. Joint subcarrier and CPU time allocation for mobile edge computing. In *Proc. IEEE Glob. Commun. Conf. (GLOBECOM) 2016*, pages 1–6. IEEE, 2016.

[462] Jing Yuan, Yu Zheng, Xing Xie, and Guangzhong Sun. Driving with knowledge from the physical world. In *Proc. of the 17th ACM SIGKDD Int. Conf. Knowl. Disc. data mining*, pages 316–324, 2011.

[463] Xiaojun Yuan, Ying-Jun Angela Zhang, Yuanming Shi, Wenjing Yan, and Hang Liu. Reconfigurable-intelligent-surface empowered wireless communications: Challenges and opportunities. *IEEE Wirel. Commun.*, 28(2):136–143, 2021.

[464] Fatih Yucel, Kemal Akkaya, and Eyuphan Bulut. Efficient and privacy preserving supplier matching for electric vehicle charging. *Ad Hoc Netw.*, 90:101730, 2019.

[465] Sameh Zakhary and Abderrahim Benslimane. On location-privacy in opportunistic mobile networks, a survey. *J. Netw. Comput. Appl.*, 103:157–170, 2018.

[466] A. A. Zazali, S. K. Subramaniam, and Z. A. Zukarnain. Flood Control Distance Vector-Hop (FCDV-Hop) localization in wireless sensor networks. *IEEE Access*, 8:206592–206613, 2020.

[467] Xuan Zha, Wei Ni, Xu Wang, Ren Ping Liu, Y Jay Guo, Xinxin Niu, and Kangfeng Zheng. The impact of link duration on the integrity of distributed mobile networks. *IEEE Trans. Inf. Forensics Secur.*, 13(9):2240–2255, 2018.

[468] Xuan Zha, Xu Wang, Wei Ni, Ren Ping Liu, Y Jay Guo, Xinxin Niu, and Kangfeng Zheng. Analytic model on data security in vanets. In *Proc. 17th Int. Symp. Commun. Inf. Technol. (ISCIT)*, pages 1–6. IEEE, 2017.

[469] Chuan Zhang, Liehuang Zhu, Chang Xu, Kashif Sharif, Kai Ding, Ximeng Liu, Xiaojiang Du, and Mohsen Guizani. TPPR: A trust-based and privacy-preserving platoon recommendation scheme in VANET. *IEEE Trans. Serv. Comput.*, 15(2):806–818, 2022.

[470] Jiaxing Zhang, Wen Xiao, Benjamin Coifman, and Jon P Mills. Vehicle tracking and speed estimation from roadside lidar. *IEEE J. Sel. Top. Appl. Earth Obs. Remote Sens.*, 13:5597–5608, 2020.

[471] Junwei Zhang, Fan Yang, Zhuo Ma, Zhuzhu Wang, Ximeng Liu, and Jianfeng Ma. A decentralized location privacy-preserving spatial crowdsourcing for internet of vehicles. *IEEE Trans. Intell. Transp. Syst.*, 22(4):2299–2313, 2020.

[472] P Zhang, S Shu, and M Zhou. An online fault detection method based on SVM-grid for cloud computing systems. *IEEE/CAA J. Automat. Sinica*, 5(2):445–456, 2018.

[473] Shaobo Zhang, Xiong Li, Zhiyuan Tan, Tao Peng, and Guojun Wang. A caching and spatial k-anonymity driven privacy enhancement scheme in continuous location-based services. *Future Gener. Comput. Syst.*, 94:40–50, 2019.

[474] Shibin Zhang, Yan Chang, Lili Yan, Zhiwei Sheng, Fan Yang, Guihua Han, Yuanyuan Huang, and Jinyue Xia. Quantum communication networks and trust management: a survey. *Comput. Mater. Continua*, 61(3):1145–1174, 2019.

[475] Yan Zhang and Dengguo Feng. Efficient attribute proofs in anonymous credential using attribute-based cryptography. In *Int. Conf. Inf. Commun. Secur. (ICICS)*, pages 408–415. Springer, 2012.

[476] Yong-Bing Zhang, Qiu-Yu Zhang, Zong-Yi Li, Yan Yan, and Mo-Yi Zhang. A k-anonymous location privacy protection method of dummy based on geographical semantics. *Int. J. Netw. Secur.*, 21(6):937–946, 2019.

[477] Yunlei Zhang, Xiaolin Gong, Kaihua Liu, and Shuai Zhang. Localization and tracking of an indoor autonomous vehicle based on the phase difference of passive UHF RFID signals. *Sensors*, 21(9):3286, 2021.

[478] Zusheng Zhang, Xufei Mao, Kunxiao Zhou, and Huaqiang Yuan. Collaborative sensing-based parking tracking system with wireless magnetic sensor network. *IEEE Sens. J.*, 20(9):4859–4867, 2020.

[479] Haiyan Zhao, Jing Yan, Xiaoyuan Luo, and Xinping Gua. Privacy preserving solution for the asynchronous localization of underwater sensor networks. *IEEE/CAA J. Autom. Sin.*, 7(6):1511–1527, 2020.

[480] Liang Zhao, Weiliang Zhao, Ammar Hawbani, Ahmed Y Al-Dubai, Geyong Min, Albert Y Zomaya, and Changqing Gong. Novel online sequential learning-based adaptive routing for edge software-defined vehicular networks. *IEEE Trans. Wirel. Commun.*, 20(5):2991–3004, 2020.

[481] Qingchuan Zhao, Haohuang Wen, Zhiqiang Lin, Dong Xuan, and Ness Shroff. On the accuracy of measured proximity of bluetooth-based contact tracing apps. In *Proc. Int. Conf. Security Priv. Commun. Syst*, pages 49–60. Springer, 2020.

[482] Yang Zhao, Jun Zhao, Mengmeng Yang, Teng Wang, Ning Wang, Lingjuan Lyu, Dusit Niyato, and Kwok-Yan Lam. Local differential privacy-based federated learning for internet of things. *IEEE Internet Things J.*, 8(11):8836–8853, 2020.

[483] Ziwen Zhao, Ayong Ye, Lingyu Meng, and Qiang Zhang. Pseudonym changing for vehicles in VANETs: A game-theoretic analysis based approach. In *Proc. Int. Conf. Networking Network Appl. (NaNA) 2019*, pages 70–74. IEEE, 2019.

[484] Haibin Zheng, Qianhong Wu, Bo Qin, Lin Zhong, Shuangyu He, and Jianwei Liu. Linkable group signature for auditing anonymous communication. In *Proc. Australasian Conf. Inf. Secur. Privacy (ACISP)*, pages 304–321. Springer, 2018.

[485] Huadi Zheng and Haibo Hu. Missile: A system of mobile inertial sensor-based sensitive indoor location eavesdropping. *IEEE Trans. Inf. Forens. Secur.*, 15:3137–3151, 2019.

[486] Yu Zheng, Quannan Li, Yukun Chen, Xing Xie, and Wei-Ying Ma. Understanding mobility based on GPS data. In *Proc. of the 10th Int. Conf. Ubi. comput.*, pages 312–321, 2008.

[487] Hong Zhong, Jingyue Ni, Jie Cui, Jing Zhang, and Lu Liu. Personalized location privacy protection based on vehicle movement regularity in vehicular networks. *IEEE Syst. J.*, 2021.

[488] Jun Zhou, Zhenfu Cao, Xiaolei Dong, and Athanasios V Vasilakos. Security and privacy for cloud-based IoT: Challenges. *IEEE Commun. Mag.*, 55(1):26–33, 2017.

[489] Jun Zhou, Zhenfu Cao, Zhan Qin, Xiaolei Dong, and Kui Ren. LPPA: Lightweight privacy-preserving authentication from efficient multi-key secure outsourced computation for location-based services in VANETs. *IEEE Trans. Inf. Forens. Secur.*, 15:420–434, 2019.

[490] Lu Zhou, Le Yu, Suguo Du, Haojin Zhu, and Cailian Chen. Achieving differentially private location privacy in edge-assistant connected vehicles. *IEEE Internet Things J.*, 6(3):4472–4481, 2018.

[491] Sophia L Zhou, Xianhan Jia, Samuel P Skinner, William Yang, and Isabelle Claude. Lessons on mobile apps for COVID-19 from China. *J. Saf. Sci. Res.*, 2(2):40–49, 2021.

[492] L. Zhu, M. Li, Z. Zhang, and Z. Qin. ASAP: An anonymous smart-parking and payment scheme in vehicular networks. *IEEE Trans. Dependable Secure Comput.*, 17(4):703–715, 2020.

[493] Liehuang Zhu, Chuan Zhang, Chang Xu, Xiaojiang Du, Nadra Guizani, and Kashif Sharif. Traffic monitoring in self-organizing VANETs: A privacy-preserving mechanism for speed collection and analysis. *IEEE Wirel.*, 26(6):18–23, 2019.

[494] Liping Zhu, Shang Wang, Chengyang Li, and Zhongguo Yang. License plate recognition in urban road based on vehicle tracking and result integration. *J. Intell. Syst.*, 29(1):1587–1597, 2020.

[495] Yishi Zhu, Bomin Mao, Yuichi Kawamoto, and Nei Kato. Intelligent reflecting surface-aided vehicular networks toward 6G: Vision, proposal, and future directions. *IEEE Veh. Technol. Mag.*, 16(4):48–56, 2021.

[496] Mayra Zurbarán, Karen Avila, Pedro Wightman, and Michael Fernandez. Near-rand: Noise-based location obfuscation based on random neighboring points. *IEEE Latin America Trans.*, 13(11):3661–3667, 2015.

Index

Note: **Bold** page numbers refer to **tables** and *Italic* page numbers refer to *figures*.